LIFE ON THE OCTOPUS FARM

LIFE ON THE OCTOPUS FARM

The Ethics and Future of
Growing the World's Most
Intelligent Invertebrate

RICHARD SCHWEID

THE UNIVERSITY OF NORTH CAROLINA PRESS
CHAPEL HILL

This book was published with the assistance of the
H. Eugene and Lillian Lehman Fund of the University of North Carolina Press.

Designed by Lindsay Starr
Set in Quadraat Pro and Halis Rounded
by codeMantra

Cover art © Nokhoog / Adobe Stock.

All translations from Spanish are by the author unless otherwise noted.

Library of Congress Cataloging-in-Publication Data
Names: Schweid, Richard, 1946– author
Title: Life on the octopus farm : the ethics and future of growing
the world's most intelligent invertebrate / Richard Schweid.
Description: Chapel Hill : University of North Carolina Press, [2026] |
Includes bibliographical references and index.
Identifiers: LCCN 2026013398 | ISBN 9781469696874 cloth alk. paper |
ISBN 9781469696881 paperback alk. paper | ISBN 9781469686943 epub |
ISBN 9781469696898 pdf
Subjects: LCSH: Octopus fisheries—Economic aspects | Octopus fisheries—
Moral and ethical aspects | Octopuses—Effect of human beings on |
Food preferences—Moral and ethical aspects | Sustainable aquaculture
Classification: LCC SH374 .S48 2026
LC record available at https://lccn.loc.gov/2026013398

For product safety concerns under the European Union's General Product Safety
Regulation (EU GPSR), please contact gpsr@mare-nostrum.co.uk or write to the
University of North Carolina Press and Mare Nostrum Group B.V., Doelen 72,
4831 GR Breda, The Netherlands.

DEDICATED TO JEFFREY JUDD IN GRATITUDE
FOR MORE THAN A HALF CENTURY OF
FRIENDSHIP AND BROTHERHOOD.

Contents

LIFE ON THE OCTOPUS FARM

INTRODUCTION

O N   A   W I N D Y, overcast day in March 2010, I was standing beside a large, outdoor tank at a marine research station in Sisal, a fishing village on the northern Yucatán coast beside the Gulf of Mexico. The small *Octopus maya* being netted from it represented the first time in the whole long history of the world that farmed octopuses (the generally accepted plural spelling) were being commercially harvested. I was there because I was searching for the answers to a couple of questions: How do you farm an octopus, and is it ethical to do so? In order to satisfy my curiosity, I had taken a job at one of the few places in the world where people were trying to farm octopuses so I could see for myself what was involved in octopod aquaculture. I had a hard time imagining how such a strange and remarkable animal might be tended.

Fish farming was easier to understand. Dig a pond, fill it with water, stock it with fish, and toss in food on a regular basis. Let them reproduce and take the fish out when they reach edible size. People are believed to have been practicing this kind of aquaculture for thousands of years. Historians place the first fish farms in China, as far back as thirteen thousand years ago, and the first written manual for farming fish—carp in this case—was written by Fan Li, a Chinese politician turned fish farmer. In his manual, written about 475 BCE, he recommended catching small carp and raising them in ponds. He assured his readers that there was a lot of money to be made doing it. People with the same notion have been farming fish ever since.

Fish were kept and grown in ponds by the Egyptians and the Romans, and in many other places down through history, but never before the twenty-first century have so many farmed fish fed so many people. Today, more than three billion people rely on fish and seafood as their main sources of protein, according to the *Financial Times*.[1] Global landings of wild fish have not increased for over a quarter century. Aquaculture is the world's fastest-growing source of animal protein, and for more than a decade now, over half the fish and seafood consumed in the world is farmed.[2] In 2022, for the first time in recorded history, aquaculture produced more than capture fisheries.[3] This pattern is expected to continue as aquaculture techniques become more sophisticated and wild stocks grow ever scarcer—and more expensive. Aquaculture around the world is currently estimated to be worth about $300 billion, and around twenty million people are estimated to work in fish farming.

Okay, so I understood that lots of different kinds of fish were being cultivated. But octopus? That odd-looking, intelligent, solitary invertebrate, with its eight arms and an ability to escape from just about any enclosure? It did not strike me as surprising that no one had yet put farmed octopus on the market. What surprised me was that some people around the globe were trying to do so and spending a lot of time and money on their efforts. But, despite all that time and money, and what I saw happening on that long gone 2010 day in Sisal, by 2025 nowhere in the world were octopuses being successfully farmed, although people were still trying.

Some popular species of fish and seafood have proven relatively easy to grow. Aquacultured shrimp, sea bream, turbot, catfish, oysters, mussels, tilapia, trout, and salmon are all produced in industrial quantities around the world and sold on the international market. They each have created their own problems to be resolved, often regarding environmental degradation, disease, or hybridization of their conspecies in the wild, but they all have managed to reach the global market in large numbers.

Other species, although consumed in quantity from the wild, have not proven so well-adapted to farming, even with many millions spent on trying and a huge potential market beckoning. For instance, large amounts of time and money have been spent trying to farm eels, widely consumed across Europe and Asia, but decades of research have yet to succeed in breeding eels in commercial numbers that survive to adulthood. At the same time, the population of eels in the wild is rapidly diminishing, and if people are going to have eels to eat in the future, time is growing short.

The octopus is a different story. Researchers in at least three countries have successfully closed the reproductive cycle of octopus with the goal of farming them. Female octopuses have produced eggs in captivity, which in turn hatched and grew into adult octopuses that reproduced, a cycle that could continue for generation after generation. By 2024, this had been accomplished in Mexico, Japan, and Spain, three countries where the octopus is widely appreciated as food and where its market value is high enough to make the idea of farming them attractive.

In each of these countries, the capacity to reproduce this most remarkable animal in captivity was achieved only after decades of intense research. It is not particularly difficult to have a female octopus produce a clutch of eggs and thereby obtain a crop of hatched paralarvae, but after that, things become much more challenging. The paralarvae of *Octopus vulgaris*, the common octopus, which is the species most widely consumed in the US and Europe, and those of *Octopus sinensis*, the similar octopus eaten by the Japanese, drift in the water column for a month or two after hatching, before finally settling to the bottom where they will live out their lives. What to feed the paralarvae while they are drifting, how to keep them alive, and how to help them thrive when they finally do settle have proven much more complicated than originally anticipated.

It took decades to begin to find solutions, and only when that was achieved did people begin to imagine successful octopus farms. At various moments during those decades of research, entities in one or another of those three countries announced that they were on the verge of bringing octopuses to market. In no time, they assured the public, they would be producing a steady, reliable supply to meet a rising demand. But all their predictions failed to materialize. Demand continued to rise, but there was nowhere that octopuses were being farmed to meet it.

The efforts to add aquacultured octopus to the seafood sections of supermarkets or to the menus of upscale restaurants were by no means universally well-received. Over the decades it took to learn how to farm octopuses, the well-being of the animals that humans were already farming—chickens, pigs, sheep, and cows—began to receive increased attention from the general public and to play a greater role in consumer decisions. More and more people became concerned about how the meat they ate reached their tables and what kinds of conditions had produced it.

This means that even if the day dawns when the biological difficulties of commercially farming octopus are overcome, ethical debates, politics, and public relations are going to come into play. The projected aquaculture of octopus has a hard sell in front of it, with a number of problematic aspects, ranging from the close confinement of what are solitary, intelligent animals, to the methods that would be used to slaughter them. Powerful animal rights lobbies around the world have responded to the idea of octopus aquaculture by threatening lawsuits and organizing large protests, and they are prepared to swing into action whenever and wherever a successful octopus farm should appear.

Those opposed to octopus aquaculture often cite a 2021 review of more than three hundred articles by researchers at the London School of Economics, which concluded that octopuses are sentient beings, capable of experiencing pain, distress, and happiness, and that high-welfare farming of them would be impossible. The report determined that the requirements of octopus life in the wild cannot be successfully replicated in aquaculture.[4]

That has not stopped people from trying, because whoever succeeds in bringing farmed octopus to market stands to make money. A lot of money. Over the past decade, sales of wild-caught octopuses have skyrocketed; they are staples wherever sushi is served and stars on the menus of some of the world's most expensive restaurants. While wild octopus numbers have not yet fallen so low as to put the most commonly eaten species in danger of extinction, their numbers are steadily declining in many places.

High global demand for octopus has led to overfishing of what were once huge populations. Worldwide landings doubled from 2.2 million metric tons in 1980 to more than four million in 2010, according to Food and Agriculture Organization (FAO) statistics.[5] In the United States, consumption of octopus has increased fourfold in the past four decades, and it continues to climb. As a result, commercial fishing vessels have overfished many of the places where octopuses once thrived. In many locations, what used to be a modest, productive, local fishery has been transformed into an industrial model, driving octopus populations into a downward spiral.

In spite of the concerns raised by animal rights activists, research designed to create a viable, commercial octopus aquaculture is moving ahead in a number of places around the globe. It appears highly possible that farmed octopus will eventually become available to consumers. Among the things we shall need to consider when pondering the ethics of farming octopus—and whether we want to buy it—are why we eat what we eat; how we make those choices; and the increasing human demand for wild-caught octopuses. We also need to consider the animal itself, so different from us, yet so similar, with its surprising level of cognition and its remarkable physiology.

Marine biologists learned a lot about octopuses over the twentieth century, and in many ways the more we know the more extraordinary the animal seems. It has three hearts, blue blood, and nine brains, a central brain, and primitive brains in each of its eight arms, with each arm packed with neurons and able to act independently of the others. All its arms can be regenerated if they are lost. The octopus has an extraordinary ability to instantaneously camouflage itself by adopting its color and skin pattern to its surroundings, and it possesses an uncanny cognitive capacity which far outstrips that of other invertebrates. Octopuses have the largest brain-to-body-mass ratio of any invertebrate. Their brains are capable of the kinds of learning that researchers long associated with higher orders of mammals, even though features of human brains, like the cortex or thalamus, are not present in an octopus's brain. "You just look at its brain and it almost looks like an alien type brain with all these different lobes. Nothing that looks familiar," one researcher told an interviewer.[6]

Octopuses are one of four animals belonging to a class of coleoid mollusks called cephalopods—the other three are squid, cuttlefish, and nautiluses. This means that octopuses are much more closely related to a clam than to a human, making their cognitive abilities all the more astonishing. Many people, including a number of marine biologists and researchers, hold that not only does it have nine brains, but it has a mind, something that we generally associate with higher orders of mammals like dogs, cats, primates, and humans.

The octopus is a solitary, short-lived marine animal that likes to keep close to home, to hunt, to eat, and to reproduce. Why would it need a mind anything remotely like ours? To all appearances it is about as far removed from *Homo sapiens* as any animal could be. But appearances can be deceiving, and, in fact, we share a number of important evolutionary traits with octopuses. Essentially, we both belong to species without shells, claws, or fangs to defend ourselves, and we have had to develop strategies for keeping ourselves safe. Both humans and octopuses have remarkable survival skills and capacities for adaptation, and that's where a mind *can* come in handy.

Whether they have minds or not, when many people around the world think of octopuses, they think of food. Octopus is a texture/taste food, highly pleasing to both senses, and it is not difficult to prepare. It has no fat or bones and is rich in healthy proteins and omega-3. The author Eugene Linden summed it up well in his book *The Octopus and the Orangutan* when he wrote, "The difficulties of imagining an intelligent mollusk are embodied in octopus websites, many of which alternate stories of octopus intelligence with recipes for cooking the animal."[7]

In this book, I consider the question of farmed octopus by journeying to the places where its aquaculture is most advanced, as well as by reviewing what we know about the amazing biology and cognition of this most unusual invertebrate. I also explore the question of why we eat what we eat; whether octopus could, or should, represent an important healthy and sustainable addition to our larders; and whether octopus might contribute to increasing food security for people around the world.

When I first started seriously searching for a place I might go to learn about octopus farming back in 2010, an online search quickly showed me that there were not many places where research was underway, but I did find contact details for what seemed to be the only functioning octopus farm on earth, located on Mexico's Yucatán Peninsula. From my home in Barcelona, Spain, I sent an email to the farm's director, Dr. Carlos Rosas, and he wrote back, encouraging me to come work with them for a while and see for myself what they were doing.

SISAL, MEXICO

DISASTER struck the octopus farm in the fall of 2010, not long after I left my job there. The farm was located at the marine research center belonging to the Autonomous University of Mexico (UNAM), just outside of Sisal, set on a barren patch of sandy scrubland beside the gulf. On the night of September 28, the lone security guard at the center had just finished making his midnight rounds. His job called for him to leave the guard shack once an hour, walk the grounds, and make sure that all was in order. He had noticed smoke toward the back of the installation, near the sea, but he assumed it was some locals burning trash on the beach, as they often did late in the evening, because Sisal's trash collection fell somewhere between sporadic and nonexistent.

The smoke got progressively blacker and denser, and the guard went to investigate. A long Quonset hut, which everyone called the Igloo, and which housed the ten long plastic tanks holding the farm's baby octopuses, was ablaze. Even if the guard had immediately spotted it, Sisal had no fire trucks—the closest were in the town of Hunucmá, almost fifteen miles inland, and they would not have arrived in time.

When the guard finally realized the Igloo was on fire, the conflagration was well advanced. The long hut didn't burn easily, but once the flames caught, the fire was intense. The baby octopuses didn't stand a chance. Some thirteen thousand of them died in the tanks that night, no doubt trying to make it to shelter under the whelk shells scattered on the floors of each tank—the only homes they had ever known—as the Igloo burned down to its cement foundation.

Many people in Sisal immediately suspected that the fire was arson. Octopus production and sales had been picking up, beginning with the harvest I had seen that spring, and there were plans to expand the farm to the land next door. This was opposed by a group of villagers who coveted the land earmarked for the octopus farm, and numerous Sisaleños were certain someone from that group had set the fire. An investigation was opened.

Just a few months before that blaze, in the same Igloo, I had handled a baby octopus for the first time in my life. Or it had handled me. Less than an inch long, it had not yet developed the personality or intelligence for which octopuses are justly famous. Although tiny, the two-week-old animal was clutching the tip of my forefinger with all the might of its eight miniature arms. I dipped my finger in the twenty-four-by-twelve-foot tank of circulating salt water where the baby lived and wiggled the digit back and forth until the infant finally relaxed its suckers and jetted away to hide under the nearest whelk shell, abandoning my fingertip and leaving the thinnest filament of ink in the water behind it. Even at such a young age, an octopus still produces ink to divert a predator's attention and, hopefully, cover its escape.

This was one of three babies that I had inadvertently siphoned out of tank number one while I was cleaning it that morning. My first task on the job each day at the octopus farm was cleaning out leftover bits of food paste from the tanks. The four women of the Los Moluscos del Mayab (Mayan Mollusks) cooperative, in Sisal, were kind enough to let me join them at their daily labors. In 2010, these women were, as it happened, the only people in the whole world who were cultivating octopuses in a commercially feasible manner, and I wanted to see how they were doing it.

The octopus they were farming was the Gulf of Mexico's *Octopus maya*, also known as "four-eyed octopus," because of the two spots on its head which look like an extra pair of eyes. It was first recognized as a distinct species in 1966. Worldwide, most of the cephalopods consumed by humans are squid, which account for 80 percent of cephalopod consumption, with 10 percent more being octopus, and the other 10 percent are cuttlefish. There are some three hundred known species in the genera of octopus around the globe, but few are eaten by humans.[1]

O. maya is only a little smaller than the two most widely consumed species, the common octopus—*O. vulgaris*—fished by Europeans in the Atlantic and Mediterranean, and *O. sinensis*, which the Japanese catch off their coasts. While *O. maya* is closely related to the other two, and tastes just about the same, it is different in a number of ways. For farming purposes, the most important distinction is that the Mayan octopus, as soon as it hatches, goes directly to live and feed on the bottom, immediately beginning its benthic life, and does not go through a planktonic stage of drifting in the water column.

How the drifting *O. vulgaris* paralarvae thrive, and what they eat, has been the bane of octopus researchers for decades. Despite substantial amounts of time and money spent in places like Spain and Japan on efforts to farm them, *O. vulgaris* researchers have had great difficulty discovering how to overcome an unacceptably high paralarval mortality. The fact that *O. maya* does not go through this planktonic stage makes it much easier to cultivate.

When we spoke about an internship at the octopus farm, the university's Dr. Carlos Rosas Vázquez, the marine biologist who was a cofounder of the research station and who directed the Laboratory of Experimental Octopus Production (better known as the Octopus Project), told me the farm was highly labor intensive, and he didn't expect there would be a problem finding work for me with the cooperative. Particularly since I wasn't asking to be paid. "Highly labor intensive" was an understatement. Daily, the four women and I sat on folding chairs around a plastic tabletop and put thousands of dabs of paste made from ground-up crabmeat, squid, and vitamins into small clam shells by hand, one smear of paste to a shell, arranging them on cafeteria trays to store in a refrigerator until feeding time. Each shell was about the size of a nickel, and Sisal's long, deserted beach was covered with them. The paste was concocted onsite in a different building by Carlos Rosas's graduate students. Twice a day, morning and afternoon inside the Igloo, we broadcast the shells carrying the paste into the plastic tanks full of baby octopuses.

The first task each morning was to remove the previous day's thousands of shells, now empty, then to siphon out all the residual particles of paste that had fallen out of the shells and was left on the bottom of the tanks. The empty clam shells we took out with small dip nets. The waste was siphoned out with a length of hose, a process initiated with one end of the hose in a tank and the other end in your mouth, sucking until saltwater came up, at which point—suction established—you took the hose out of your mouth and put it in a bucket at your feet while using the end in the tank to vacuum. If you siphoned correctly, you got the suction going without swallowing more than a few drops of water. My first days on the job I drank what seemed like gallons of the Gulf of Mexico, which is where the water came from that recirculated in the tanks. I provided plenty of slapstick entertainment for my colleagues as I choked and spluttered.

Inevitably, during siphoning, the hose would suck up some babies caught outside the whelk shells where they lived. The bucket into which the hose emptied had a screen on one side so that water flowed out to a drain in the floor to recirculate, but debris, residual food, and baby octopuses did not. Fortunately, the experience of being vacuumed out of a tank into a bucket at a tender age did not seem to affect them. After siphoning a tank, we went carefully through what was left in the bucket, discarding debris and food remains and returning the babies to the water, where they jetted vigorously away toward the closest shell.

Each octopus was valuable, because only 50 percent of the babies would survive to weigh a gram after some three weeks, when they would be transferred to outdoor tanks. In the tanks, they would be fed on spider crabs, and virtually all of them would reach "gourmet" table size in a couple of months. Each would have cost about two dollars to raise, and they would be worth up to ten dollars on a restaurant menu. To the women of the Mayan Mollusks cooperative—who had spent the past five years figuring out how to farm *Octopus maya* under the direction of Carlos Rosas, and had done so for the first three of those five years without any salary whatsoever—every animal deserved the best care they could give it. Flushing one down the drain with the waste did not sit well.

．．．．．．．．

Every afternoon when I got off work at the farm, I walked back to the house where I was staying and stretched out in the hammock inside my upstairs rented room, too tired to do more than watch a small brown lizard crawl across the ceiling as I picked at the dried paste on my palm, peeling it off little by little. I was always tuckered out. After years of sitting in front of a computer every day, I'd forgotten the ways other kinds of work can tire a person. Downstairs, the family that owned the house would have a talk show radio program playing loud enough to wake the dead. Someone hammered constantly next door, a parrot in a cage squawked in the backyard beneath my window, a rooster crowed intermittently down the street, the wind off the gulf howled around the house, and a generator frequently ran at top volume somewhere close by. I would swing gently back and forth in the hammock and review my gripes with the place: The television didn't work, the family was noisy, the toilet only flushed irregularly, there was no lock on the door to my room, and a few mosquitoes got in every night to ruin my rest.

On the other hand, the rent for the room was only eighty-five dollars a month, while across the road that ran in front of the house was a mangrove lagoon full of flamingos, and a hundred yards behind the house was a deserted beach and the Gulf of Mexico as far as the eye could see. Food was cheap and tasty. The tiny fridge in my room was a little noisy, but it kept the beer cold.

And, I was learning how octopuses are farmed.

On Friday, March 12, 2010, Mexico was, yet again, in the headlines of media around the world, and the news was no better than usual: Narco-terrorists in Ciudad Juárez had murdered three US citizens, and travel advisories were issued for three Mexican states by the US State Department. At least thirteen people were killed, some of them beheaded, around a popular beach resort in Acapulco, just as college students from the United States began arriving for spring break.

It was that same Friday afternoon when I watched that first crop of octopuses netted from their tanks. Sisal was crime free, as usual, but the first-ever-in-the-history-of-the-world octopus harvest was very big news to the members of the Moluscos del Mayab. Given the nature of the global media, it's not surprising that this historic moment in aquaculture went unreported, but it was the culmination of five years of hard work for the cooperative.

Some four hundred million pounds of wild octopus is bought and sold around the globe each year. Plenty of economic incentive exists to farm them, but it has proven difficult. On that day in 2010 when I stood with Carlos Rosas watching a couple of dozen hundred-gram octopuses get netted from their tanks, he told me he was convinced that Sisal's Octopus Project had the problems solved. He was fifty years old and a cofounder of the Autonomous University of Mexico (UNAM) marine biology station in 2003. He was of medium height with light skin, thinning brown hair, wire-rimmed spectacles, and his graying moustache and beard were neatly trimmed. Carlos Rosas was excited standing by that tank, almost euphoric, as the small octopuses were dipped out to be killed and frozen. He believed that from this point on everything would begin to move quicker.

"This is the first time we've had a harvest with a systematic production," he told me. "Before this it has been sporadic. If we can produce these, we can produce tens of thousands. I'm very moved and very proud of the people who work here. Sometimes a researcher like myself can dream of things outside of reality, but here in Sisal the people do not have that luxury. We've shown today that our project can work in the real world."

The plan for the Octopus Project called for the cooperative to purchase the land adjoining UNAM with grants and business development loans, with the goal of turning it into an octopus farm. It would have thirty outdoor tanks for reproduction and fattening, indoor facilities for incubation and pre-fattening, a processing plant, and a small restaurant. All of this would happen within the next two years, according to Carlos Rosas's best-case scenario, so that by the end of 2012, Mayan Mollusks would be a thriving business, regularly selling and shipping small, tender *Octopus maya* to high-end restaurants with "gourmet" cuisines across Mexico. If it happened, it would also represent the creation of a viable octopus aquaculture, the first of its kind in the world. Patent revenues for the nutritional and infrastructural solutions devised in the process would be divided equally between the university and the cooperative, he told me, as we watched the octopuses get harvested.

"I believe it's going to happen," said María Virginia "Genny" Narcisa Uicab Novelo, forty-four, who along with Silvia del Carmen Canul Pardenilla was one of the two original cooperative members still on the job. "I've got five kids, two are married, and three are still in school. My youngest daughter's fourteen. I expect to see my kids inherit a share in an octopus farm."

That would be a welcome addition to what most people in Sisal usually inherited, which ranged from nothing to a little piece of land with a rudimentary concrete block house on it, perhaps an aluminum jonboat, the knowledge of how to fish, hunt, and prepare what was caught and killed, and the ability to extract sustenance from the place where they lived—a body of knowledge that had accumulated over generations. The village's privileged location between the Gulf of Mexico and a vast mangrove swamp meant that despite their deep poverty, Sisaleños were unlikely to go hungry. Food could be extracted from both the sea and the ciénega—the swamp—for those who knew what they were doing. That knowledge would put food on the table from day to day, season after season, year after year, but it did not go far when trying to meet unexpected expenses. And while it might have been a backwater, Sisal was only an hour's ride in a packed, rickety, rattling van to the big city of Mérida, twenty-five miles inland. Sisaleños were frequently confronted by the same problems as the rest of us in this twenty-first-century world: debt, substance abuse, daughters turning up pregnant, sons in trouble, sickness, teeth rotting, and all the other misfortunes that nip at peoples' heels anywhere they are trying to get through the day.

Sisal only had about 1,600 residents, and many of them were distant cousins. Families knew everything about one another. Many Sisaleños, including both Silvia Canul and Genny Narcisa, could trace at least part of their families back to the Mayan culture of precolonial times, before the Spaniards arrived. The *Calkini Codex* mentions Sisal; the codex was one of only four Mayan volumes that survived the book burnings carried out by the Spanish bishops when they brought their "one true faith" to the Yucatán in the sixteenth century. The codex records that in the eleventh century a Mayan priest named Ah Kin Canul kept four boats in the Sisal harbor, manned by slaves who fished the rich surrounding waters. One of the things they were likely to have brought back from their fishing trips was octopus. Fray Diego de Landa, who wrote the first Spanish account of Mayan life on the Yucatán Peninsula between 1562 and 1573, wrote, "There is very good octopus on the coast of Campeche," about a hundred miles southwest of Sisal.[2]

Sisal, with its deepwater channel connecting the Gulf of Mexico to a sheltered port, was likely to have been an important part of Mayan economic life in the region. By the time Spaniards arrived, the Maya had lived and governed in the midst of the vast and fertile jungles of Southern Mexico and Central America for millennia. Walking the jungle paths and climbing the pyramids at one of the many leftovers of that Mayan civilization, in the hundreds of miles between Tikal in Guatemala and the Mayapan ruins some twenty-five miles south of Mérida, is enough to transmit an unmistakable sense of ancient urban organization in the Mayan world. It is immediately clear that the people who inhabited these places lived, died, worked, played, and worshipped for many centuries in the midst of a highly organized society, which covered a lot of territory. "This was no priest-plus-peasant society, but a vastly stratified, cosmopolitan culture," writes the University of Pennsylvania archaeologist William Coe, who supervised some of the most important Mayan digs.[3]

Sisal's importance continued after the Spaniards arrived, serving from the mid-sixteenth century as a port and gateway to Mérida and the rest of the peninsula for both people and goods. This was only enhanced when Sisal was named Yucatán's official port in 1810. The original customs house is still standing, a block from the beach, with a lovely arched portico on Sisal's main street, constructed around an ample, open, interior patio. As befitted an important port, around 1562 a dirt road was built across the mangrove swamp dividing Sisal from the rest of the Yucatán, and that road connected it to the city of Mérida, or Tiho, as the Maya named it. There was also a large international shipping trade between places like New Orleans to the north or Havana to the east. Sisal prospered.

This long run as an important port came to an end in 1870, when Mexico's president, Benito Juárez García, announced that the nearby coastal village of Progreso would be equipped as the region's new official port and would henceforth be the main shipping point for Mérida. The decision came not because Progreso was a couple of miles closer to Mérida but in retaliation for the warm welcome given by Sisal to the Empress Charlotte, who visited in 1865, during the three years when she and her husband, Maximillian, ruled Mexico for France. Subsequently, Benito Juárez and his supporters ousted them and reclaimed Mexico for Mexicans, executing Maximillian, leaving Charlotte to live the rest of her life exiled in Belgium, and cutting Sisal off at its economic knees.

Between the mid-nineteenth and mid-twentieth centuries, the village retooled its economy, basing it on the henequen fiber, also called "sisal" in English. It was used to make rope and cordage around the world, and it was extracted from the arms of the agave cactus, which can grow as tall as an adult person. Nowhere in the world does it grow better than in this part of the Yucatán Peninsula. In 1868, two years before Sisal lost hegemony to Progreso, some four million pounds of sisal were shipped out of its port. This number plunged rapidly when the port was moved to Progreso, but the sisal industry still managed to contribute substantially to the economy of the village and its surroundings. The agave cacti were grown on vast plantations called haciendas—some 250 of them in the state of Yucatán alone—where brutal working conditions and abysmally low wages were the order of the day, but work was still available even after the port collapsed.

Despite the 1917 Mexican Revolution, land reform did not greatly affect the sisal-based economy. Most of the haciendas continued to grow agave and to use Indigenous labor to strip out its fiber. The local economy maintained the huge gulf between landowner and laborer. Darker-skinned Mayans continued to do the dirty work on the plantations, which were owned by the light-skinned descendants of Spaniards who came to the New World to strike it rich, and they did so, constituting Mérida's wealthy class. Initially, many of the haciendas had served as second homes away from the city, but as the henequen boom developed, the landowners moved to the plantations to oversee their operations. They built opulent homes from which to conduct business, while Maya grew the agave and stripped the sisal from its leaves for a miserable wage and long hours.

It was not until President Lázaro Cárdenas's agrarian reforms of 1937 instituted control of the land by cooperatives of workers that the production of sisal finally began to benefit the people who did the work. As part of Mexico's widespread collectivization process, Mexican farmland was reorganized as ejidos, owned by the members of local cooperatives who farmed them. In 2025, despite a decades-long return toward privatization, some 50 percent of Mexico's farmland is still owned by cooperative members. In the case of workers on the henequen plantations, it did not do them a lot of good: The age of plastics followed the Second World War, with the development of strong, cheap, nylon rope and cord. The market for sisal virtually disappeared.

This history is legible in the twenty-five miles between Mérida and Sisal. Much of the land is covered in low scrub, spindly young trees and bushes, and overgrown fields where agave cactuses the height of a person once stretched to the horizon—land that has reverted to wilderness over the past fifty years. Here and there one comes across what's left of the haciendas: huge stone houses crumbling to ruins and open to the weather, vines crawling over the moldering remains of elegant rooms that once held all the things money could buy. Machinery for stripping the fiber from the cactus lies rusting in outbuildings.

Progreso has grown into a thriving port, business center, home to a colony of North American retirees, and a popular tourist destination with cruise ships docking throughout the year. Sisal has slipped into obscurity, and stayed there. By the early 1950s, it was basically a dusty, forgotten fishing village, pretty much untouched by tourism or any other source of ready cash. Pictures of Sisal's main street in 1959, archived in Mérida's Center for the Support of Yucatán Research, looked virtually identical to Sisal's main street when I got there in 2010, give or take a cantina or two. In other 1959 photos, the beach stretched to the lighthouse, the sands as empty as they were when I walked them fifty years later.

· · · · · · · ·

When I first started working at the farm, I would walk to and from the job along a dusty road, which turned into sand as it left the village and crossed the dunes toward UNAM's marine research campus. One day, Doña Genny asked me what route I was walking to work. When I said I used the road and then the sandy trail, she told me to try walking along the beach instead, which began just in the back of the campus. "It's a lot nicer."

She was right. I saw pelicans diving; dolphins breaching; iguanas with their peculiar rolling gait making their way across the sand; soaring frigate birds with their sleek black bodies and forked tails slashing across the blue sky; and flamboyances of flamingos flying, long, pink bodies stretched out like arrows, elongated necks straining forward, startling black and red wings spread overhead in flight on their way to the mangrove swamp. I was always the only person on the beach.

Occasionally, I would walk back to my room, rest for a while, then rouse myself from the hammock and go back to the beach for the sunset. I saw some spectacular ones, large, fluffy clouds brightening to rose-pink and flaming orange as the sun went down behind them into the blue-green expanse of the Gulf of Mexico. One evening I was sitting on a cement bench on the town pier, looking west across an empty mile of beach to the lighthouse, astounded by the brilliant colors on the horizon as the sun went down. As often happened, I was the only person on the pier. A strong north wind, a norte, was blowing into shore, and the air was scrubbed and fresh. I couldn't help but wonder how we could not love the world that allows us to thrive for a while amidst such wonder.

Suddenly, in the middle distance, three guys appeared on the beach walking toward the pier. As they drew closer, I looked at them and got a jolt. A bad jolt. One of them, hard-faced and in his late thirties, was holding a two-by-four down by his side, his large hand wrapped around it in a way that threatened serious damage, and the other two, who appeared younger, actually had their faces covered, one with a black and white bandana tied across mouth and nose, and the other with his head deep in the recesses of a hooded sweatshirt. They looked truly menacing. I let my glance slide right off them, and when they stepped up from the beach onto the pier and walked past me into town, I saw that the man carrying the two-by-four had a machete stuck, handle down, in the back pocket of his jeans. If these guys are on their way to visit somebody, I thought, I'm certainly glad it's not me. They walked away down the deserted, dusty main street. I stayed seated and watched the sun go down, no one else in sight.

· · · · · · · · ·

When Benito Juárez transferred the official port to Progreso in 1870, Sisal's lovely customs house was left to slowly crumble behind thick, locked, wooden doors. It was in the ruins of this abandoned stone building that the seeds of the Mayan Mollusks cooperative were planted. Literally. In 2003, a small group of women from Sisal—among whom were Silvia Canul and Genny Narcisa—decided to clear the garbage and rubble out of the historic building and plant a garden in the rich, old soil of the patio, where they would grow things to sell. They planted radishes, parsley, cilantro, eggplants, sweet peppers, and hot peppers, selling their herbs and vegetables in the village.

Doña Silvia Canul was the leader of the initiative from its beginnings. Leadership suited her. She was short and square, a hard worker, always on the go, moving through her days with determination and perseverance, organizing one task after another. She was president of the cooperative and the driving force at the octopus farm. It was Doña Silvia who told us what to do each day and when we should do it.

"My husband, Don Antonio, is a fisherman," she told me, as we sat around the white plastic tabletop putting paste in the tiny clam shells, in the storeroom behind the large, vaulted Igloo where the tanks were housed. "We needed more than just his income, and the kids were old enough not to need minding. A group of us decided it wasn't right that such an old historic building as the customs house was just a garbage dump. We went in and cleared it out, cleaned it up, and planted vegetables and herbs in the court-yard. We also instituted cinema nights in the building and put in a small library.

"Then we met Dr. Carlos. He came to see us and said he could give us better work, and he explained his project. He convinced us. The first three years we worked with the octopuses we didn't get paid a salary. Not anything. We made a little money selling the grown octopus, and that was it."

Sisal was one of those places in the world where it was possible to get along with very little cash money. The fishing was good, when the weather over the Gulf of Mexico permitted. The ciénega, the mangrove swamp that passed behind the town, was part of a sixty-mile-long wetland, which provided waterfowl and blue crab for the table during their seasons.

Octopus season in the Gulf of Mexico ran from August through December, and there was no shortage of ready buyers. In 2009, some twenty-two million pounds of *Octopus maya* were landed in the Yucatán, and there were twenty octopus processing plants in Progreso alone, feeding the global market during the season. In Sisal, hundreds of fishing boats were registered in the port, almost all of them open, twenty-six-foot fiberglass launches with forty-horsepower outboard engines on the back.

Standing by the breachway channel that led to the port on any given morning during octopus season, when the norte was not blowing too hard and waves were not too high, a long line of those open launches would go past on their way out to the gulf. Each carried a small dory that would be put overboard when the fishing grounds were reached, and each boat had a handful of long poles sticking out over the bow or the stern. The pole is called a *jimba* and is used to fish for octopus in a way that is only practiced widely in the Yucatán and which octopus fishers will tell you was invented in the coastal town of Campeche and subsequently adopted by everyone up and down the Yucatecan coast. Each jimba is a bamboo pole some twenty feet long, which has a number of lines suspended from it. A crab is tied on the end of each line and lowered to the bottom. An octopus latches on and refuses to turn its prey loose even as it is hauled aboard a fishing boat.

Unlike more sophisticated, industrial forms of fishing, in which everything beneath the water is raked up by trawlers and separated on board, fishing with jimbas is ecologically sound in that it leaves the bottom intact. Females do not eat while they are tending their eggs, and therefore they will not be attracted to the bait, plus there is no bycatch since no hooks or nets are used and only octopuses are caught.

For all the octopus fished there, Sisal did not have even one processing plant. The fishermen's cooperative bought all the catch. Refrigerated trucks from Progreso were on hand to buy the octopuses from the cooperative when the fishermen came in and carry them to a plant. Progreso is only twenty miles west along the coast, but the road that used to unite the two villages was destroyed in 1988 by the devastating Hurricane Gilbert. While every governor of the State of Yucatán since then has made a campaign promise to repair the road, it has yet to happen, so transport has to go inland from Sisal to Mérida on one highway and then out to Progreso on another, more than doubling the length of the trip.

Of course, it could be worse. It was not until the early 1950s that the road from Hunucmá to Mérida was paved, and it took almost another decade before Sisal was connected by an asphalt highway to Hunucmá. Until that road was laid down, a trip to Hunucmá was infrequent, and many people from Sisal visited Mérida only a few times in their lives. People in the village were isolated and generally self-sufficient, with little recourse to the outside world.

By 2010 that had changed, but not so much. The traffic between Hunucmá, Mérida, and Sisal was usually light, and the village was still pretty much its own world. While Sisaleños generally had enough at hand to live on, most folks still were not able to do more than just get along day to day—eat, sleep, and, with luck, have some kind of work to go to. Day or night, walking down the dusty streets, it was easy to look in open doors or windows and see how few things people owned. The small houses were basically bare of furniture. In the living room were often a television and some hammocks, one for each member of a family. Here is where people stretched out, swung gently, conversed, read, watched television, and slept. After the midday meal, when the sun was broiling hot, people often laid on the cool tile floors, doors open to catch any breeze. If there was other furniture it was likely to be some red or blue plastic cantina chairs, with the Coca-Cola logo or Sol beer written across the back, and maybe a plastic table. A bed was not something usually found in these homes. Sisaleños prefer hammocks, and someone who slept in a bed had probably lived elsewhere for a while, most likely north of the border. Returnees from the United States often brought a predilection for beds, and probably the bed itself, back with them.

In Sisal, not much opportunity existed for saving money or putting a little something aside. The occasional construction job might turn up, building a second home close to the beach for a middle-class family from the big city of Mérida, or cooking and cleaning for those families when they were in residence. The village's mercantile opportunities were restricted to a number of small stores offering limited groceries, cigarettes, and beer, along with a handful of bars and restaurants. Sales were cash only in these places; they were not equipped to accept credit cards. For those who needed to buy more than just basic staples, a trip to the district capital of Hunucmá, nearly fifteen miles inland, was required.

In addition to the inconveniences of living in a village, Sisal also offered small-town rewards. The pace of life was generally unhurried, and the first word Sisaleños invariably used to describe their village to me in 2010 was tranquilo. People left their bicycles unchained outside their homes overnight, and break-ins, or any crimes other than an occasional drunk and disorderly, were rare.

Nevertheless, life in Sisal was far from easy. Storms could come up quickly in this part of the gulf, and every year octopus fishers drowned. Like fishers in many places, many of them did not know how to swim. Nor did they carry lifejackets in their boats. They reasoned that if they went overboard in a sudden storm out in the gulf, they were likely to be goners, too far from any place to be reached by swimming.

The risk of drowning was far from the only downside to life in Sisal. The hottest times of the year were virtually unbearable outside at midday. Mosquitoes were omnipresent, and the mangrove swamp had its miasmas. The mosquitoes transmitted dengue fever—otherwise known as breakbone fever—and it seemed like everyone you asked had suffered through its awful fevers, aches, and pains at one time or another; in addition, cholera was no stranger to this region. Further, the fact was, it was just plain tough to be poor in Mexico, and the State of Yucatán was one of the poorest in the country. Sisal was predominantly and devoutly Roman Catholic. An image of the Virgin of Guadalupe was on walls everywhere, as she was across Mexico. Families were large, and money was scarce. Men often had drinking problems—Sisal, with a population of sixteen hundred, has had as many as three different Alcoholics Anonymous groups meeting at a given time.

Doña Genny Narcisa had dark brown skin, jet-black hair, and was shy until she knew you, then she was warm and thoughtful. She grew up with ten brothers and sisters, and they were all raised by her single mother who worked every waking minute of her life, according to Genny, at whatever she could find: making tortillas to sell door-to-door, weaving and repairing hammocks, or cleaning other people's houses. She put her kids to work, too; there was no other choice if the family was going to eat—the children's father had left after the birth of the last child and contributed nothing. Doña Genny told me that she was working in a little store by the time she was ten, combining the job with school attendance. At fourteen, she was married ("in church") to seventeen-year-old Julio Ernesto Sierra Delgado, known to everyone, including his wife, as "Baby," a nickname reflecting his and his wife's tender ages when they pledged their troth.

"Don Baby had to come and court me in the store while I was working," she told me. "That was the only time he could see me. We girls couldn't go out unless my mother was going somewhere and we could all go together. Girls didn't go out by themselves, much less with boys."

Baby, forty-seven, was one of two men who were members of the six-person Mayan Mollusks cooperative, and the other was Silvia Canul's husband Don Antonio Cob Reyes. Don Baby had been married to Doña Genny for thirty years. He was solidly built with curly, uncombed black hair, slightly graying. He had powerful hands and the muscular forearms of someone who had hauled things out of the water all his life. At some point, the forefinger on his left hand was crushed and badly repaired, and it was permanently stuck straight out; he could not bend it.

Baby was a quiet, self-contained, and self-reliant person. Indoors and out he wore a blue baseball cap, on the back of which white letters read: "A. A. Grupo Sisal." It seemed like such a contradiction to advertise membership in Alcoholics Anonymous that I had to ask about it. His dark, battered, strong face creased in a snaggle-toothed grin. "We're a group of crazies," he laughed. "Anyhow, you can't keep anything a secret in this village. I've been sober for twenty-three years."

He worked every day at the farm, tending to the larger octopuses in the tanks outside, but he still got up at 4:30 a.m. on mornings when weather and season permitted and went out to haul and reset his nets for the sardine-like charales or fish for octopus before coming to work. If the octopus farm thrives, he told me, he'd like nothing better than to give up fishing. It was hard work, and the pay was low, with the middleman who bought his catch making most of the money.

In late 2003, when Carlos Rosas cofounded UNAM's marine research station with his colleague Adolpho Sánchez Zamora, it didn't have much immediate impact on the lives of Sisal's residents, but from the beginning the professor hoped that the station would contribute to both scientific knowledge and the well-being of the village where it was located. Once he joined forces with the women in 2005, he helped them to navigate the tricky waters of Mexican bureaucracy in forming their cooperative. In 2008, the group received its first grant from a government foundation to encourage small cooperatives, and each member was making three thousand pesos ($250) a month in 2010.

"I was formed politically in Mexico City in the seventies and eighties, and I do feel that if I have a job in which I receive a salary paid for by the public with their taxes, then the least I can do is give some of that back," Carlos Rosas told me.

"In Mexico, not many scientists do this. Not because they don't want to, necessarily, but because the traditional path of research and publication takes them in a different direction. Octopus is now giving me the chance to work with people and return a little of what I have received, and I'm very grateful for that."

Carlos Rosas was the sort of teenager who did not give his parents much trouble. He was doing marine biology before he even knew that's what you called it. He built his first aquarium in his parents' garage, and with a group of friends from the neighborhood raised fish to sell to other kids and, eventually, to commercial aquariums. "And there was also Cousteau. In the seventies, Jacques Cousteau had an excellent program, and I never missed one. I would watch Jacques Cousteau, and then I would scuba dive under the dining room table."

In addition to his attraction to marine biology, Rosas had a strong sense of social justice from an early age; both of these were honed during his university years. His doctorate research was on crustaceans, and he spent the first twenty years of his career working with shrimp, which became Mexico's most important aquaculture, generating large amounts of money. He was still working with crustaceans when he came to Sisal to cofound the teaching and research station. In exchange for approving the construction of UNAM's branch at Sisal, the Yucatán state government asked that part of the research done there focus on things that were of local interest, and gave them a list.

"*Octopus maya* was on the list," Rosas recalled. "Although I'd spent twenty years working with shrimp and crabs, it seemed like an attractive challenge. I was getting a little tired, academically, of crustaceans. Switching to octopus was like falling in love again. I didn't know anything about *Octopus maya*, but I felt like research with them could have some practical, utilitarian use. Basic science is fine, but it often lacks applicability in the short run. We designed the Octopus Project with this in mind."

From the beginning, he was able to combine his love of pure research with the possibility of developing a new source of income for the community, one that could be managed by women. He wanted the Octopus Project to demonstrate that scientists working in their laboratories—and people laboring at whatever nonscientific things they did—had a lot to offer each other, and if they worked together it could make life easier for both groups. By the end of 2004, he had completed his extensive studies of *O. maya* to the point that he felt ready to draw up a research and marketing plan. It was then that another professor at the center told him about the group of local women who had restored the customs house and were growing plants there. It resonated with him, so he sought them out.

"When I heard about these women I thought: If they're doing this, they can do anything. They must be just as crazy as I am. And I went to talk to them. I told them I wanted to do an experiment that involved getting wild octopuses, putting them in tanks, and seeing if they grow or not, seeing if we can find something to give them to eat that they like. I offered the ladies a simple deal: 'The product is yours—I just want the data.'"

The farm started with three fiberglass outdoor tanks, each stocked with one hundred kilos of octopus bought from local fishermen. The octopuses they bought each weighed around three hundred grams, and when they grew to a kilogram, the cooperative sold them and replaced them in the tanks with new three-hundred-gram animals. What they found was that the weight gain could happen within a month, much faster than Carlos Rosas had anticipated. "I was absolutely astounded at how quickly they grew, and of course the ladies were very happy because they had quite a few octopuses to sell," he said. "When the octopus season ended, they stayed on with us helping construct the whole system and participating in experiments, doing trial runs of the equipment.

"That's how our relationship began, and when octopus season came around again, they filled the tanks with octopus to fatten. Then I proposed that we would develop the technology for them to use in a year-round business, and that's when they went through the legal process of becoming a cooperative. It's a long, ongoing process, but we hope that in two more years they can have their own land and build an octopus farm on it."

Cooperatives are a common feature of Mexico's economic life and have been since precolonial times, perhaps for millennia. Economic life among the Maya, described by Fray Diego de Landa in his sixteenth-century description of the Indigenous population found by the Spanish colonists, made it clear that the cooperative spirit was alive and well there many centuries ago:

> The Indians have the generous custom of helping one another in all their work. At the time of sowing, those who do not have any people of their own to carry out the task gather in groups of twenty, sometimes more or sometimes fewer, and all together complete the tilling . . . and do not stop this work until all have been helped. Land, at the moment, is common, so that the first to occupy it owns it.[4]

This cooperative spirit persists throughout Mexico. More than three million campesinos still live and work in more than thirty thousand ejidos, and often they are people whose parents and grandparents did likewise. In 1992, under the administration of Carlos Salinas de Gortari, a law was passed that—for the first time—allowed an ejido to award its members individual titles to their *parcelas*, not merely usufruct rights. Ejidatarios could now choose to rent, sell, or mortgage their parcelas. Additionally, ejidatarios did not need to work their lands to maintain ownership of them, and they could enter into partnerships with private entrepreneurs.

When the law was passed, some observers predicted that many ejidatarios would sell their parcelas into private hands for the first fistful of pesos offered. They predicted that the days of cooperatively owned ejidos were numbered and that their acreage and membership would dwindle away to virtually nothing. This has, apparently, not proven to be the case. While some ejidos have been bought in part or whole by national or multinational corporate agricultural interests, the members of many other ejidos have voted not to privatize ownership but to retain the ejido's cooperative structure. In 2011, some 5.6 million acres of farmland on the Yucatán Peninsula were still administered by ejidos.[5]

The terrible violence of the drug cartels has also affected ejidos in numerous parts of the country. When cannabis or poppies for opium produce many times more wealth than corn or beans, land can take on a new value. With a lot of money in play, some ejidos have been subsumed into cartels, and others have resisted, often with violent consequences. Regardless of the eventual fate of the ejido system, Sisal already had its share of cooperatives, and it was not difficult for residents to accept the creation of yet another at the UNAM installation. Nor was it difficult for the cooperative's members to work together under that structure. What was difficult was trying to make it pay something. Anything.

The octopus farm's first years proved difficult for both scientist and Moluscos del Mayab members. The captive octopuses were fed a diet of crab mixed with fish waste—heads and carcasses. The women would bicycle into town from the research station each day, buy crabs and collect fish waste from the small fresh-fish market and the village's handful of cafés, take everything back to the farm, cut it up, and feed it to the octopuses. Bicycling the two miles out from the center of Sisal to the station on a sandy dirt road, transporting a load of fish carcasses for no pay under a brutal sun, required a lot of faith.

Carlos Rosas had convinced the cooperative members that it was worth the effort, but in those early years their fisher husbands were skeptical of the plan to farm octopus. They were perfectly willing to sell the adult octopuses they caught to the research station, which they did, but along with most of the other people in Sisal they were dubious about the newcomers from the university with their odd, ivory-tower ideas.

"In the beginning, the men didn't believe it would work," recalled Silvia Canul. "It was only us women who were doing it at the start. When I'd get home and tell Antonio about what we were doing and how fast the octopuses were growing, he would just brush it aside, or say we were crazy.

"Dr. Carlos always said women were preferable because they are careful and have more patience than men, which is what you need to do this work. It was also Dr. Carlos who suggested to Don Antonio and Don Baby that they should come and see for themselves, and he invited them to join the cooperative we were forming and become part of it. Up until then, we women did everything from cutting the PVC pipe to installing the tanks; we did everything from zero. It wasn't easy. We were just fishermen's wives. We didn't know much. We learned on the job with Dr. Carlos."

As it turned out, the scientist did not have an easy learning curve either. He knew how hard the women were working, but in order to take advantage of the fact that newly hatched *Octopus maya* go directly to the bottom without a paralarval planktonic stage, the eggs—each about the size of a large grain of rice—first had to hatch. And Carlos Rosas had the devil of a time finding a way to make that happen. "It was easy to get eggs, but they just wouldn't hatch. They're porous, full of nutrients, and bacteria attack them. Sometimes the female would eat them. We tried everything and nothing worked.

A long string of disappointments. For me, that was the most discouraging time. We'd get eggs, but we couldn't get octopuses."

In the wild, female octopuses produce clusters of eggs strung together by a chitinous material. In *Octopus vulgaris*, the eggs are small but numerous, perhaps a hundred thousand of them. This correlates with the fact that the paralarvae will be drifting in the water column for weeks, easy prey to any of the many other marine animals with a taste for them. Not many will survive to adulthood. The eggs of *Octopus maya* are larger, and when the paralarvae hatch they go straight to the bottom and find shelter. However, each female will only produce about sixteen hundred of them, and they are still appetizing to many predators.

In both species, the mother will hang her clusters of eggs somewhere—from the ceiling of a rock cave, or on the wall of her den, or some other convenient spot in deep shadow. Then, in both species, the female stops eating and spends her last weeks tending to the eggs, siphoning water over them, keeping them clean. Once they hatch she will die, passing through a terminal phase called senescence, which often bears considerable resemblance to dementia in humans. The starving, wasting female octopus will wander outside her den, aimless and listless, not eating, with a seeming unawareness of what's around her, without any apparent desire to live. A few species, like the small, lesser striped octopus *Octopus chierchiae*, found in Nicaragua, will produce eggs more than once, and a female of this species has produced eggs up to three different times under the surprised gaze of researchers. But females of *O. vulgaris*, *O. maya*, and *O. sinensis* produce fertilized eggs only once.

Females will keep the eggs spotless by jetting water over them constantly during the day and brushing them clean with a gentle arm. It was not until Carlos Rosas began to pay close attention to replicating this behavior that he was able to hatch eggs. "Finally, after a year, we put the eggs in a tank with a jet of water to keep them clean, all in low light. Ten eggs hatched, and one of those survived. Our first. We called him Octavio. One morning I came in to check him out as I did every day, and he had climbed out of his tank and died. It was our fault; we didn't feed him well enough, or he wouldn't have gone looking for something to eat. Nevertheless, with Octavio we made an important advance. I began to think we could do it."

For much of his career as a marine biologist, Carlos Rosas followed the standard scientific path of teaching, research, and publishing. He was still doing so, to some extent, when I met him. But increasingly, what was important to him as a scientist was applying in the field the research he was doing in the laboratory. "For us, the important thing is that the technology of everything we do has a solid scientific base, but at the same time we want to develop the technology with the cooperative in a way that allows us to rapidly take what we learn in the laboratory and apply it to production."

When he tried something new and it worked, he did not wait to do it again or run controlled experiments to validate his results and write them up. What he did was put whatever worked into immediate effect on the Octopus Project, and move to the next problem. There was never a shortage of problems to solve, but the number of available graduate and post-graduate students at the UNAM research station increased year by year. He had a growing group of lab workers to take on the experiments he suggested. The students had the pleasure of seeing their results immediately applied.

When the UNAM station was inaugurated in June 2004, it had four tenured professors, seven teaching assistants, forty students, a secretary, and a security guard. By 2010, the station had a hundred students, twenty tenured professors, fourteen teaching assistants, new classrooms, and a well-equipped laboratory. And the most advanced octopus aquaculture in the world. The farm was still not a moneymaking operation when I arrived in 2010, but it was getting closer, and Carlos Rosas attributed much of that progress to the way research and production teams worked together. "The change in how we structured research here is why we have the success we have today," he told me. "It permitted us in very little time to reach an ample level of knowledge about the octopus and its production. What we have here has clearly become a technology and not just an experiment."

The system in place at the farm when I arrived was developed with hard work at each stage. The outdoor tanks held the mature octopuses with a variety of large conch shells and ceramic cylinders for shelter. In these tanks, male and females cohabited and went through their strange copulation. The male octopus has an arm called a hectocotylus, the third arm from the right eye, and an octopus can be sexed by inspecting this arm to determine whether it has the peripheral canal that carries the packets of sperm, the spermataphores. The hectocotylus's specific function is copulation and the transfer of spermataphores to the female's mantle cavity near the oviduct.

Once the transfer is accomplished, no further relationship between the female and the male takes place. The female retains the spermatozoa until her eggs mature, at which point she fertilizes them. In fact, in a bizarre iteration of polyamory, she can retain spermatozoa from matings with more than one male, and eggs from the same female are frequently found to have been fertilized by different males. Once she produces the eggs, she will devote the rest of her short life to tending them and will die when they hatch. The male may outlive her by a few months and pass spermatophores to other females, but the males also die after a year—or two at most—of life.

After spending a couple of weeks with males in the outdoor tanks, the adult females were assumed to have been inseminated. When one of them stopped eating, it meant she was about to begin laying eggs. She was brought indoors to the reproduction area and placed in an individual tank. The reproduction area was kept in near darkness, which was how gravid females liked it, and the permanent twilight and constant murmur of circulating seawater made it a tranquil environment for the expectant mothers. From the time she was brought inside, it might be two to four weeks before a female produced eggs, and once she began it could take five days to lay them all and hang them on the wall of her shelter. Once the clusters of eggs were extruded, they were collected by a farm worker and put in small incubation tanks where they were kept constantly clean. In another six to eight weeks, the eggs would hatch, and the quarter-inch-long hatchlings would be transferred to the care of the cooperative, in what was called the pre-fattening area, which was where I worked.

The hatchlings were fed with artemia for the first week of their lives, that odd little brine shrimp, a species of which is also known as "sea monkey." Artemia were the same animals that, as a kid, I sent away for with a coupon from the back of a comic book and a dollar bill. An envelope arrived inside a box, and when I opened it up and dumped the powder into a bowl of water, each grain of powder suddenly became a live, swimming "sea monkey." Dried artemia can stay alive, in stasis, for up to two years without oxygen, ready to respire when water is added. They reproduce in large numbers and are an economical feed for fish farming. They are also a favorite of octopuses, and the newly hatched denizens of our pre-fattening tank consumed them in great quantities. However, trial and error at the farm had proved that artemia do not entirely fulfill the nutritional requirements of growing octopuses, and after a week of sea monkeys the babies were moved up to the paste in clam shells.

To me, the squid/shrimp paste had the unappetizing look and feel of something formulated from thick, pureed oatmeal mixed with petroleum jelly. However, the babies took to it right away. *Octopus maya* will eat a wide variety of living creatures, but one of its favorite foods is crab. When an octopus captures a crab, and has it beneath its mantle, it injects a venom into one of the crab's eyes, which relaxes the captive's musculature and makes it easier to open and reach the meat inside.

The venom present in an octopus's saliva is being researched to discover what beneficial properties it might offer the pharmaceutical industry. One study published in 2021 found that the venom of the Australian southern sand octopus had a peptide capable of inhibiting proliferation of the deadliest type of melanoma cells—skin cancer—in mice. This peptide was baptized Octopep-1, and in 2023 research was ongoing. Biochemical vendors were offering it to laboratories at prices upwards of $360 for a gram.[6] The effect of octopus venom on a crab is almost instantaneous, at which point the octopus finishes opening the shell and scoops out every last morsel of meat, leaving behind an empty half shell.

Baby and juvenile octopus that do not get enough to eat inevitably turn on their tankmates; cannibalism has been a persistent problem for would-be octopus farmers. Data shows, Carlos Rosas said, that the level of cannibalism is directly related to whether the babies are eating well. What constitutes eating enough for a baby *O. maya* turns out to be a lot. They have big appetites. No one went home until the clamshells and the paste they contained were tossed into the tanks for the afternoon feeding, and a couple of people pulled duty to feed the juveniles on Saturdays and Sundays. This was a farm, and like animal farms everywhere it was a seven-day-a-week operation.

As important as it was to make sure the octopuses had plenty to eat, I was grateful to find that my colleagues from the cooperative made sure we ate well, too. A good appetite was something the farm workers shared with their charges. Once the morning's initial chores were performed—tanks cleaned of old shells, siphoned, and replenished with new food—we gathered around the plastic tabletop in the storeroom to put the paste for the afternoon feeding in shells, but first we had breakfast. This usually included weak instant coffee, cola, soft "French" bread, cheese, pastries, tamales, empanadas, *panuchos*, chitlins, some blood sausage, sliced ham, a pile of tortillas, and whatever was left over from lunch the day before.

Ah, lunch. It was a daily lesson in Yucatecan home cooking, with the main dish prepared by Doña Genny over a charcoal fire outside, behind the storeroom. What with going to school and working in a little store, she never had a chance to learn how to cook from her mother, she told me, and she had to teach herself. She had certainly done a fine job of it. Each day, she spent the last part of the morning out back, preparing lunch for the Octopus Project workers. We ate at the same table where we put paste in the shells. The storeroom had a loading-dock door at the back, open to the outside during the day. Not surprisingly, given the quality of the meals Doña Genny put on the table, a number of people from other parts of the UNAM station would find a reason to pass by the storeroom looking for something right around lunchtime, or they just happened to be walking by and glanced in to see what was on the table. "Hm?" they would raise their eyebrows, asking permission, peel off a tortilla from the pile on the table, fill it with whatever we were eating, and go on their way.

It took weeks before I could convince Genny to let me give her the money for a day's lunch ingredients, but after I continued to pester her about the cost of the delicious meals I was eating, she told me okay—it was my turn to pay. I gave her eleven dollars in the morning, and at midday it fed a table of eight people on plenty of short ribs grilled over the fire, served with a salad of chopped lettuce in a fresh, tasty lime juice dressing, rice with carrots chopped up in it, sliced avocados, tortillas, black beans called *frijoles negros*, and a big bowl of limes to squeeze on the ribs.

Also on the table every day to liven up lunch was Don Baby's hot sauce, which he made from fresh, chopped habanero peppers, salt, and lime juice, all mixed together in a Philadelphia cream cheese box. A shelf in the bodega held spices, including a bottle of hot sauce called Fuego Maya, and a jar of powdered habanero peppers, but Baby still deemed it necessary to make his own. I was surprised the cream cheese box didn't smolder and immolate. His sauce was incendiary; it lit up the tongue.

In 2010, the hottest pepper in the world, judged by a universally accepted standard called the Scoville scale, was the bhut jolokia from Assam in northeast India. The stubby habanero from Central America was the second hottest pepper regularly consumed by human beings. A few drops of Don Baby's sauce was enough to add a wealth of flavor and bite to a whole plate of rice. People occasionally came by the storeroom with their lunch in Tupperware that they had heated in some nearby microwave, or a tortilla full of something they had brought from home, to ask, "Don Baby, got any of that hot sauce?," spooning just a bit out of the cream cheese box on the table. No one used more of the hot sauce than Baby himself, and by the time lunch was over he was always sweating profusely. And he was happy. He was what my father used to call a trencherman, someone who does real justice to a meal. He only had three visible teeth in his mouth, but they were enough so that his wife Doña Genny said, laughing, that when Don Baby was done eating supper she didn't even have anything left to give to the dogs. It was truly a pleasure to share the midday table with him.

When the farm began back in 2005, the women went home for lunch. They worked from 7 to 11 a.m., then went home and did housework, fixed lunch for their kids, and were back at the octopus farm to work from 3 to

7 p.m. Once the farm's scope went from experimental to commercial, that routine no longer functioned—it simply did not give them the time they needed to get a full day's work done. "In those early days we didn't have so many little octopuses, and we could go home for lunch, but now we have to eat lunch here every day. There's just not any other way to do it," Doña Silvia told me one afternoon, as we sat around the table after the midday meal, putting more paste in shells.

I silently thanked the gods that be for that, particularly as my stomach was full of that day's lunch of coots. Seven species of ducks are hunted in the ciénega during the winter months, according to the father of the cooperative's youngest member, Bianca Daniela (Dani) Hernández Hernández, a twenty-five-year-old single mother of two young daughters. He had shot these the day before from his blind in the mangrove swamp, and Doña Genny grilled them over a bucket of charcoal in the back and served them in their own broth with lime, cilantro, chopped onion, rice, black beans, and plenty of tortillas, with Baby's hot sauce on the side. A couple of drops turned the broth into a tasty three-alarm fire. We had eaten well. The meal had been cleared, the table wiped down, and once again we were seated around it, putting paste in shells.

"We're working all day long now, and it's hard," Doña Silvia said. "My daughter's in high school and we have different schedules, so I hardly get to see her. Fortunately, my mother lives next door and she helps out with cooking for us. Some evenings when I get home, all I want to do is head for my hammock and lie down. My son works in Mérida and when he's home on the weekends I have to come out here on Saturday or Sunday to feed the babies. 'Mama,' he tells me, 'you love those animals more than me.' But we've already sacrificed so much for the project; we can't leave it now when we're just starting to see some return."

Time was helped to pass quickly on the job by a constant stream of conversation around the table. No one wore earphones to tune into their private music, and there was not even a radio in the storeroom. People carried cell phones, but they were only for emergencies, and most days went by without anyone receiving a call. The women talked about the same things people talk about all over the world—food, the weather, who was sick and who was well, who was having a romance, or a fight with neighbors—and they did so for hours on end. They all knew the histories of each other's families, which made it easier to dish dirt at the table, and sexual innuendo was highly appreciated. One day an enterprising local man passed by selling fresh eggs and Don Baby bought half a dozen, which he put in the refrigerator and forgot to take home that night. The next day, everyone who passed through the storeroom heard about how Don Baby had forgotten his *huevos*, and every time the story was repeated it engendered another round of laughter at the double meaning of eggs.

"It's hard work, but the time goes by quickly because everyone tries to enjoy themselves," Dani Hernández told me. She lived with her family behind a patio that served as a restaurant called El Niño de Atocha. Dani's mother and an older sister prepared and served food there, and had been doing so for the past ten years. They were excellent cooks. What's more, the sign outside that read "COCINA ECONOMICA" was telling the whole truth and nothing but the truth. It really was economical—three dollars for a full and delicious supper. Dani was attractive and lively, and she had the standard short, square physique of Yucatecan women, with long black hair and dark, shining eyes. She had come back to Sisal and her parents' home after living long enough with a young guy in Mérida to have two daughters and for him to skip out on the family without a backward glance. She was not happy to be living with her parents, she told me, but at the moment she had no alternative. She initially found work doing maintenance at the UNAM station, and then she had joined the Octopus Project. Her parents looked after the little girls during the day in their house behind the restaurant.

Dani's younger brother Iván, nineteen, also lived with Dani, her daughters, and his parents. He came to work at the UNAM campus as a maintenance assistant after his sister got her job. He was tall, thin, and taciturn, with a rare bright smile. On weekends, I often saw him riding through the village streets on a lovely, tall, dappled gray mare. He looked like a kind of gangly kid until you saw him on his horse. He rode with his back straight, head held high, smiling broadly, fairly beaming with pride, and there was often a girl seated in front of him whom he reached around to hold the reins. He fed, watered, and brushed the mare every afternoon after work. His father had advised him not to buy her, he told me. "He said I didn't realize how much responsibility a horse would be, but I realized it, and I wanted to do it anyhow."

Iván's mother threw her son a nineteenth birthday party and invited all his coworkers on the Octopus Project. She hired a *trova* group to provide music and put on a lavish spread of that local scrumptious dish, the Mexican equivalent of barbecue, *cochinita pibil*, in which a pig is slaughtered, scraped, and slathered in a sauce made from the juice of bitter oranges and pimiento. Banana leaves are laid across the coals in a fire pit and the pig is set on them and left to smolder overnight. Just before eating, the meat is pulled from the bone just like pulled pork barbecue and served rolled up in corn tortillas. The group playing the ballad-ish, romantic Yucatecan trova music—a style developed in Mérida in the 1920s—consisted of two older men playing guitars and a third playing bongos. The three of them were dressed in starched white *guyaberas*, black pants, and shined black shoes. They switched off on vocals as they stood genteelly on the cement floor of the patio at El Niño de Atocha playing, while we ate and danced, and ate and danced some more.

One morning, not long after Iván's birthday party, Silvia had some bad news to share at breakfast—farm salaries were going to be late. Again. The bureaucracy involved in administering the foundation grant through the university is formidable, and while the money was certainly in the bank, its distribution was complicated, depending on more than one person. Carlos Rosas was doing everything he could to pry the money loose from the bureaucrats, but he was not having any success. This was by no means the first time salaries were late, but a weeklong Easter vacation was approaching. Without money people would be hard-pressed to enjoy it. In addition, nearly everyone had already borrowed from friends and family against the expected check, and now it was delayed.

Iván was the most visibly upset among us. It turned out that he still owed $250 for his gray mare, and if he defaulted on his fifty-dollar-a-month payment, the man would reclaim it. Iván walked around the storeroom opening drawers and slamming them shut, and for the rest of the week he wore a frown, steadily deepening as the days went by and his default drew closer. It appeared ever more likely his father's dire warnings would be vindicated. Conversation was less animated around the table, with less laughter, and dominated by speculation about whether the money would arrive.

The fourth female cooperative member, Juana (Juanita) de la Cruz Maldonado Ek, thirty-five, normally talked less than the others, maybe because she had a lion's share of immediate worries. Two years before, her husband was working on the maintenance staff at the octopus farm when he had a massive stroke. He was, finally, mostly recovered, walking and talking well, but he was not ever likely to be able to return to work. The couple had two kids. Juanita worked a second job at the Hunucmá town hall in the human resources department three afternoons a week doing paperwork. She wanted to send her sixteen-year-old son to a school in Mérida that would prepare him for the university entrance exams, but even with her two jobs she could not afford the extra $150 a month. Her family was barely getting by even when the cooperative members *did* receive their salaries.

Carlos Rosas decided to take extreme action to try to shame the responsible bureaucrat into doing his job and shake loose the funds to pay the personnel. He initiated a conference call in his office. Dani was chosen by the cooperative to represent them on the phone. Once Carlos Rosas had the bureaucrat on the line, he asked if he'd mind hearing directly from one of the affected workers, and Dani pled the case, emphasizing their hard work, patience, and the proximity of Easter. The man promised the money would be in the bank by the end of the week: Iván would keep his horse, and folks would be able to enjoy Easter. Another crisis resolved.

The farm had already been through five years' worth of crises, but that mid-March day in 2010, as octopuses were harvested from the tanks, was the first time that Carlos Rosas and the cooperative's members felt they were close to having a viable octopus farm: The animals had been successfully raised, in decent numbers, from eggs to the size they planned to market. That size was much smaller than the octopuses normally sold on the global market. This would have a number of advantages for Mayan Mollusks: A small octopus takes less time and money to farm, and it can be promoted as a gourmet item, selling for more. That way, a smaller number of animals and a more diminutive operation could turn a good profit. Competition was another consideration. There were no octopuses of this size on the market because it was illegal for fishermen in the gulf, and in most places in the world, to keep such a small octopus. Carlos Rosas was confident that because the farm was not obtaining its octopuses from the wild, it would be possible to obtain an exemption from the size limit.

A modest-scale production and an undersized octopus were what he believed was needed in Sisal. "Our production system is small. It won't allow us to produce in great quantity, but we can have a constant and standard production. We want to find a niche in the high-end market. I hope we can find a wholesaler who will take, say, one hundred *pulpitos* a week. What's important for the cooperative is to have a complete cycle between the production, the sale, and the final consumer, and to keep this constant."

Of course, in order to sell well in the gourmet market, they would have to taste good. In fact, they would have to taste great. Up until now, Carlos Rosas explained to me as we drove into Mérida from Sisal in his pickup truck, no one had tasted them. We were, in fact, on our way to an Italian restaurant, Ca' d'Oro, to try the first dozen Mayan octopuses ever raised for market—those specimens that I saw harvested from the tanks. The restaurant's chef was an acquaintance of Carlos Rosas and had volunteered to prepare them for a taste test. The professor was anxious. Anything not quite right about the taste would set the Octopus Project back for a long time.

People in Sisal eat octopuses prepared in a number of ways, including fried with rice, as ceviche, in an escabeche sauce, in a *mojo de ajo*, cooked in its own ink, or deep-fried. How they do *not* eat it is sautéed in olive oil with a bit of white wine and served under an Italian marinara sauce, but that is how the pulpitos were prepared for Carlos Rosas and me by Marcello, a chef from Naples, who came to the table to tell us how pleased he was to have had a chance to work with them. He said that he had trained in northern Italy, where octopus that size were considered a delicacy, eaten in a *frittura* or a *zupetto*. Ours were accompanied by thin slices of toast. They were marvelous, with an octopus texture and flavor, yet tender, with a faint taste of the sea. Marcello said that they also would be great with spaghetti and cherry tomatoes or chopped fine in a salad. As we ate, a smile spread across Carlos Rosas's face. "These are fantastic," he exclaimed. "This is a revelation. We've really got something here."

Carlos Rosas looked surprised when I asked about the ethics of farming an animal with the cognitive abilities of an octopus. "Although there's a very abundant [wild] fishery today, international demand continues to grow, and the fishery won't be able to grow past a certain point. There's a commercial demand. If we can help to meet that demand, at the same time as we improve life for people in these coastal communities, then we're doing a good job. We have to be careful that octopus farming doesn't become an environmental problem like shrimp farming did, which has destroyed hundreds of thousands of acres of mangrove wetlands."

As to the desirability of farming an intelligent animal, he said that octopuses have always been killed and eaten in the Yucatán. "I agree that it's a shame to eat such an intelligent animal, but a lot of octopus gets eaten in the world. Spain imports 250,000 tons a year and Japan 150,000 tons, and that doesn't count Greece, Italy, and lots of other places. Octopuses are undoubtedly intelligent, but that hasn't discouraged people from eating them. Because in addition to being smart, they're tasty. Very tasty."

Now that he was convinced the farm had a working production system in place, Carlos Rosas felt comfortable getting some marketing efforts underway. His first success was when he received an invitation to send some product to Mexico City for an event to be held just before Easter called Improvisation in the Kitchen, featuring thirty top chefs from Mexico, Spain, and Japan. It was sponsored by a company that manufactured kitchen appliances and was expressly aimed at the "gourmet" market. Among the upscale products from Mexico the chefs had to work with were six varieties of oysters, three kinds of mescal, two types of salt, and small *Octopus maya* farmed by Mayab Mollusks.

Unfortunately for us farm workers, the event was being held on a Monday in the middle of the day. This meant the octopus would have to travel to Mexico City on the 6 a.m. flight out of Mérida, which also meant that we would have to pack them for the flight on Sunday evening, each in a plastic bag of seawater with a dose of oxygen pumped in, so that Carlos Rosas could drive them to the airport early Monday morning. It was the first time live octopuses had been shipped by air from Mérida, and no one knew how many would survive the trip to Mexico City to take their already-announced turn in the lineup of delicacies.

All the workers on the Octopus Project were asked to gather at the center on Sunday at 5 p.m. to do the packing. Naturally, the only person to show up on time was the gringo. I sat on a rock by the gulf, behind the center, and waited for everyone else. Frigate birds hovered and swooped on the thermals, and brown pelicans sailed gracefully through the air, diving suddenly into the water and surfacing as they swallowed their catch. A big iguana scrambled along the rocks at the water's edge. By 6 p.m., everyone was present and hard at work, netting thirty-two octopuses out of the tanks, bagging them up, and packing them in ice chests for the next morning's flight. In a couple of hours we were done. A full moon had risen. It was Palm Sunday with a strong norte blowing. The clouds rushed across the sky, moonlight and shadows on the water. "Please utter small prayers or incantations for the survival of these octopuses, depending on your faith affiliation or lack thereof," said an anxious Carlos Rosas, as we packed up the last animals.

For whatever reason, whether the prayers had sped them safely on or we had done a good job of packing them, they all survived the dawn flight to Mexico City the next morning and made it to the tasting. As time went on, it would become clear that chefs preferred their octopuses to be somewhat larger, which would take another four to six weeks of feeding to reach, but that first shipment of small octopuses was a big hit at the Mexico City tasting. "Some were bathed in mescal. Others were combined with salt and lemon. The applause was unanimous. The little Yucatecans conquered the largest city on the planet," reported the *Diario de Yucatán* the next day.

That article made a big stir in the storeroom. Doña Silvia read its words of praise to all who passed through. It looked as if things were finally starting to look up for the cooperative's members. My time on the job was over, and it was a good time to leave. Although I was a volunteer, it was good to know that a month's salary for my colleagues—three thousand pesos ($250)—had finally wound its way through the labyrinthine bureaucracy and arrived in the coop members' pockets. In addition, Easter vacation was underway, and cars from Mérida were arriving with families who had come to pass the holiday week in their second homes. Suddenly, cars outnumbered bicycles and horses on the village streets, and every day the beaches were filled with people from the city who spent the whole day on the sand and who brought big beach umbrellas, and boom boxes, and ice chests with lots of food and beer. The atmosphere at the octopus farm, and in the village, was notably upbeat. On top of it all, the pulpitos had been lauded in Mexico City by top chefs. Life was good.

. . . .

It didn't last long. The source of the fire that killed the farm's stock of thirteen thousand baby octopuses in the fall of 2010 was never definitively uncovered. In 2023, when I went back to see what had changed in thirteen years, many in Sisal still suspected arson, but the official explanation, following a brief investigation, was a short circuit in the electrical system. Regardless of how it began, the fire was devastating, bringing an abrupt halt to the Octopus Project, just when it was finally coming together as a commercial venture.

Once the guard on duty had realized that the Igloo was burning, it was already too late to do anything, and it was clear that the fire would have to burn itself out. The guard tried to call Carlos Rosas, but he was out of town. A nongovernmental organization had invited Carlos and Antonio Cob Reyes, Doña Silvia's husband, to give a two-day workshop for a group of fishermen on the Pacific coast of Baja who were thinking about branching out into octopus farming. Carlos was to talk to them about the nutritional and environmental needs of octopuses, while Antonio would explain the hands-on labor involved in farming them. It was the first time in his life that Antonio had flown, and the two of them had been looking forward to the trip for weeks.

"We flew from Mérida to Tijuana," Carlos Rosas remembered. "They came to get us in Tijuana and took us to Ensenada by car, and the next day we went to Guerrero Negro, the town where the fishermen were. The second day of the workshop, the administrator called me from Sisal to say there had been a fire and I should come right back. I said I couldn't; I was in Guerrero Negro."

He decided not to tell Antonio about the fire until the trip was ending, so as not to interfere with the workshop, and he waited until they were in the airport about to fly back to Mérida. Antonio took the news fairly calmly, but not so his wife. When Carlos had called Doña Silvia from Guerrero Negro she was distraught and totally discouraged. For her, it was as if years of hard work were erased overnight.

Early on the morning following the fire, she had gotten a call from the research station with the bad news, and a car came to her house to bring her to the destroyed Igloo. When she saw the smoking remains, she broke into tears. "It was terrible," she remembered. "Just the morning before, we had put lots of babies in the tanks, and they were totally burned up. It was a terrible sight. I cried and cried. I couldn't sleep for nights."

It was a logical reaction, but not one shared by Carlos Rosas. "Maybe someone would say I'm insensitive, but I'm not," he told me. "Confronted with this [the fire], what would I gain by lamenting what had happened? Nothing. What I asked them to do right away was take away all the rubble, all the garbage. I wanted the cement platform cleared off so I could get to work there. I didn't have any idea of what to build, but the pump was working, and with the platform clean and some tanks there we could begin. What I wanted was to show that we were going to keep working."

The Octopus Project was brought to its knees by the fire, but it was far from finished. However, at the same time as work rebuilding the farm began, the cooperative was locked in a struggle for the land where their plans called for an expansion. It was a familiar Mexican story. The federal government had granted the land to the cooperative, but an apparently corrupt *comisario*, the elected commissioner of the village, had sold the land illegally to a businessman who wanted to excavate it, level the ground for houses, and sell the earth he dug out. The cooperative denounced the excavation to the federal environmental protection agency, which ordered work stopped in July 2010. This was followed by a visit from eight of the agency's inspectors, accompanied by four federal police officers. They were confronted by one hundred supporters of the excavation project and taken as hostages. The army had to come in and liberate them. "In the middle of all this is when the fire happened, and I believe it was an action, a sabotage," Carlos Rosas told me in 2023.

The dispute fell into the maws of Mexican bureaucracy. The land originally destined for the cooperative wound up being sold to others in Sisal who had plans to build houses for rent or sale as second homes to people from Mérida. Octopus fishers from Sisal were opposed to that sale and blocked the highway between Sisal and Hunucmá. Once again, the army and the police had to intervene. The government sent a negotiator. Among the land in question was the cooperative's. They had been granted a fifty-year federal concession on that land, but the concession had been ignored at the local level. It was finally determined that the original federal concession would be revoked and replaced by a new piece of land, larger and better situated.

Meanwhile, it took five years after the fire for a replacement of the burnt Igloo to be designed, funded, and built at the UNAM station. "Five years went by," Carlos Rosas lamented when I visited him there in late 2023. "Everything was up in the air for five years. In the end we wound up with a new building, but it was a tremendous delay."

Practically no sooner was the farm up and running again on the UNAM campus than COVID-19 shut down the world. The comisario who was at the head of the village when the plague struck was Miguel Antonio Ek Peck. He was new to the job, elected in 2018 by an extremely thin margin of twenty votes. Ek, a short, burly man in his early forties, was a supervisor at the local water treatment plant who had decided to run for office because he believed that the incumbent power structure was not doing its best for Sisal. After only a year as comisario he had to respond to COVID-19, and although he was a political novice, he didn't hesitate to make an unprecedented decision to protect Sisaleños at all costs.

"We closed Sisal down for five months," he told me. "Closed it down. For five months, no one came into Sisal and no one left except for medical emergencies. We were denounced by people in Mérida, and in Hunucmá, but there's an article in the [Mexican] constitution that says when it comes to public health the competent local health authorities have the right to take decisions if it's to safeguard the health and physical integrity of the inhabitants. That's what I did."

For five months, Sisaleños had only each other to depend on, and they could be proud of the way they did so, Miguel Ek told me. "No one showed us how to survive the pandemic. People went out to fish and shared their catch. We killed pigs and distributed the meat. Others brought us fruit and vegetables to share. The community went back to the old economy of mercantile exchange, and we did not have a single case of COVID. Not a single case."

Students who the pandemic had found at the Sisal station were locked down there for four months. Sisaleños brought them food from town, and Carlos Rosas called daily from lockdown in his home in Hunucmá. "The students couldn't leave because Sisal was closed," he told me. "They were the ones who kept the laboratory alive. In our laboratory you couldn't turn off the pumps, because if they stopped the salt accumulated until it stopped the motor. That would have been very costly, but the students kept all the system running, and when finally we could get there everything was in working order."

It was a group of fishermen from inland Hunucmá, accustomed to coming every day to work out of Sisal's port, and barred for months from entering the village, who finally convinced the comisario to open access to Sisal. That's when the first cases of COVID arrived. When he ran for a second term in 2021, Ek won by 721 votes in a near landslide.

He told me he was a wholehearted supporter of the Octopus Project. "Aquaculture could be very important to our community. For instance, this year the octopus season is not turning out as well as we had hoped, and the Moluscos del Mayab's production could help this kind of thing, and provide a consistent, steady income for some people. They have spent years trying to do that, and we can only hope they achieve it."

When I went back to Sisal in the late fall of 2023, the cooperative had been trying to make money farming octopuses for almost twenty years, and time was taking its toll. Antonio and Silvia were thirteen years older than when I had first seen them, but they were still optimistic about the farm's future. They took me to see the two-acre plot of land reserved for the new octopus farm. It was not too far from the UNAM station, close by the controversial parcel of land that had been designated for them in 2010, but this one was larger and clearly delineated as belonging to Moluscos del Mayab, with wooden signposts at the land's corners announcing them as the property owners. It was a good-sized parcel. A well had been dug and pipes laid from the gulf, and the outdoor tanks were in place. There was a cinderblock shed that would fill up with the tools and the implements of octopus farming. They showed me where a small restaurant would be built for Doña Genny to cook for workers at the farm and the paying public.

The tanks, and the pipes to pump gulf water to them, had been paid for by a $50,000 grant from a United Nations program to subsidize the development of small businesses. A second $50,000 from the same program was forthcoming in 2024. Still, there was never enough money on hand to do what needed doing. When the cooperative finally had the land and the first round of UN financing in place, a group of local investors had appeared, with a tempting offer of a substantial amount of money. "But what was the vision of the Yucatecan investors?" Carlos Rosas asked me, rhetorically. "'We'll put up the money and you work for us.' Something very good happened: The cooperative said, 'No señor, you have the money, but we have the knowledge. If you want to invest we'll all be equal members, not employees.' What happened? They lost the investors but retained the cooperative structure."

One more instance in a long history of the Maya refusing to be co-opted by a whiter, non-Mayan society. In fact, the Maya are the largest group of Indigenous peoples on the planet still speaking their own language, and they have been doing so for at least four thousand years. All of the members of the octopus farm cooperative spoke Maya with each other but were kind enough to speak in Spanish when we worked together. The Maya have never been wholly subdued and subsumed into the culture around them.

In the first half of the nineteenth century, much of the Yucatán Peninsula was divided into vast haciendas with sugarcane and henequen plantations owned by non-Mayans who conscripted the local Indigenous people to work for low wages in brutal conditions. In their book *The Mysterious Maya*, George and Gene Stuart wrote, "For most hacienda owners, a whipping post for their Indians was as necessary as watering troughs and cool shade trees for their cattle."[7]

In 1847, the Yucatecan Maya rebelled and began what came to be known as the Caste War. Over the next couple of years they were victorious enough to control nearly all the Yucatán Peninsula. They were on the verge of taking Mérida but never did so. The tide of war turned, and by 1850 the revolt was quelled, though not entirely. Isolated battles and skirmishes continued through the end of the century. The embers of Mayan resistance to non-Mayan dominance smoldered until 1994 when the Zapatista Army in the southern Mexican state of Chiapas declared itself at war with Mexico, and over the next decades it negotiated an ever-larger control and autonomy for the region's majority Mayan population.

Sisal's Indigenous population was not nearly so militant, but they were always quietly insistent and assured in their identity. They spoke their own language, as did their children, and many of the women wore the lovely white, embroidered blouses, skirts, and dresses that are traditionally Mayan. The cooperative members stressed that they were glad to have investors, but only those who would give them an equal voice in management and an equal share in the profits. Money was always scarce, but not everything was for sale.

In addition to grant monies, and revenue from octopus sales, the cooperative had other funding sources on the horizon. A Canadian adventure-tour operator who already was organizing successful trips to the Yucatán was enthusiastic about the idea of including the octopus farm as a place where her clients could pass a couple of weeks working as volunteers. She was confident she could bring a steady stream of Canadians to work on the farm by the beach, eat meals prepared by Doña Genny, and pay for the privilege.

"There are lots of times when [the project] looked dead, but even if it's not successful it doesn't matter," Carlos Rosas told me one day. "Our duty is to do this, make the discoveries, and show others that these people can do it. We keep going. Here, we're all members of the club of optimists."

Despite their optimism, Don Antonio and Doña Silvia were in their sixties. Don Baby was still going out fishing and was more than ready to help get the new site up and running, but Doña Genny had developed trouble in her legs and it was painful for her to walk even a few steps. Still, she told me, she was improving a bit week by week, and she expected to be cooking at the new installation, when the cafeteria was built, turning out her delicious meals for farm workers, tourists, and locals. Dani Hernández was also ready to get back to work, and all that was holding up progress was waiting for an audit of the cooperative's financial structure to be completed. That was followed by the second $50,000 installment from the UN, which happened in the spring of 2024, and the next phase of construction at the new site began. Octopuses were in their tanks by summer, but then came Hurricane Milton, the second most intense hurricane ever to strike the Gulf of Mexico. Tanks were destroyed and the solar system providing power was damaged. By the end of 2024, however, tanks were in place again, and the farm was up and running.

Carlos Rosas worried about the high average age of the cooperative members. "We've all lost a little strength over all these years," he told me in 2023. "We've proposed to the high school here that they include a workshop on aquaculture among their activities. The workshop will be for kids from fifteen to eighteen years old who are already used to working in restaurants or stores. That way we'll see who's really interested. Once we see that, we'll integrate them into the cooperative. At first as employees, as assistants, and in the future they'll have the possibility of becoming cooperative members."

In addition to the perpetual scarcity of funds, and too few young members, the cooperative was short of female octopuses in 2023, which were needed to begin producing eggs. Don Antonio was being paid about eighty dollars a day by the cooperative to go out and fish for them when weather permitted, in the hopes that he would bring back lots of females. In 2010, when I was working at the farm, I had pestered him to take me out so I could see how the fishers used the jimbas, and he had told me he was perfectly willing to do so, but weekend after weekend the gulf had been rough with the norte wind blowing into shore. He was not leaving port. We never did get out.

Now that I was back, thirteen years later, I continued to pester him. He was sixty-three years old but still tough, short, wiry, and muscled, still fishing for octopus, and still taking his morning runs along the beach. Running on sand kept your legs in shape, he liked to say. Finally, one Friday evening, I was notified with a phone call that he would be going out the next day, and if I wanted to come along I could meet him at the farm at 7 a.m. Of course I wanted to go, even though the prospect of getting up on Saturday morning in time to make the long walk out to the farm by 7 a.m. was a little daunting. I set my alarm and managed to fall asleep that night by 10:30. At 1:30 in the morning I was jerked wide awake by a loudspeaker out in the street, a very loud loudspeaker, broadcasting an off-key, male voice singing words I could not make out, accompanied by an out-of-tune guitar.

I waited for it to stop, hoping I would be able to get back to sleep when it did. But it didn't stop for a long while, and when it did, I heard voices murmuring, then the singing started again at the same excessive volume. I looked outside, and down the dusty street was a house with its windows lit up and a handful of people gathered beneath a huge loudspeaker, tall as a person, perched on top of a tripod set up at the edge of the street. Everyone must be mighty drunk, I reckoned, but still I figured it wouldn't last long. This was, after all, an extremely quiet village at night, one of the most tranquil places after dark I had ever lived. I was wrong about it ending soon. The singing kept on, loud enough to rattle windows, pausing regularly for some response from inside the cinderblock house, then starting up again. I couldn't make out any other words than the occasional "amor," so I guessed it was someone singing a love song, maybe serenading a beauty inside the house. I could have cared less about what it was; I just wanted it to stop so I could go back to sleep, and I grew increasingly irritated.

After an hour, I couldn't stand it. I was furious and went out on the 2:30 a.m. street, mad enough that I didn't even take the time to change out of my pajamas and house shoes. I was surprised to find a lot more people than I expected, sitting and standing in the shadows in front of the illuminated house, which I saw was full of people and also wreaths of flowers. No one seemed drunk. It was the stuff of some odd dream—myself in pajamas and slippers wandering through a group of strangers without any idea of what was going on. I approached a middle-aged man sitting in an aluminum folding chair, who looked up mildly at me. He was apparently not surprised to see a stranger, in pajamas and bedroom slippers, in front of him at 2:30 a.m. "Excuse me," I said, my voice tinged with anger. "What the hell is going on?"

"It's for the dead," he replied, nodding toward the house. "He died."

His answer rocked me back on my heels. What I had mistaken for amoroso was religioso. "Is the dead man inside?" I asked.

"No, he died exactly a year ago at this time of night, and we do this to remember him and his death," he told me in a soft voice as he considered my pajamas. "We'll be done in half an hour."

I thanked him, chastened and touched. No one is likely to gather and sing for me a year after I go, wherever and whenever that should happen. I went back to my room. Sure enough, after another half hour quiet returned to the street, and I slept until six o'clock.

I was early to the farm and waited for Antonio, watching a steady parade of boats with their jimbas leaving port for the open water, while in the air above them egrets, herons, frigate birds, and pelicans swooped. I must have watched fifty identical boats head out to a relatively calm gulf in single file through the short channel from Sisal's port: twenty-six-foot fiberglass, wide-beamed boats with forty-horsepower outboard motors on behind, each with two or three crew aboard, carrying long poles that reached beyond the bow or the stern. Also on board was a small boat or two, light, bathtub-size dories big enough for one person and a pair of jimbas, which would be unloaded with a crew member off the bigger boat when the fishing grounds were reached.

We can assume that people have been passing through this channel in boats on their way to open water and a day's fishing for millennia. Because octopuses are boneless invertebrates, they leave behind very little archeological evidence, but the remains of vertebrate fish show that Mayan people living near the coast have always consumed food from the sea. Mayan wooden dugout canoes have been found, which date back more than a thousand years. The canoes are thought to have been made by hollowing out a single tree trunk. Fishing was done by nets, traps, or hooks. Even if octopus was not the preferred catch, it is not hard to imagine that when a fisher in pre-Columbian Sisal caught one, it did not get thrown back. It is highly likely that octopuses have formed a part of the local diet since these areas were settled.

When Antonio showed up, the first thing we did was buy a bagful of spider crabs at the port for bait. We loaded them on his launch with a pair of jimbas, fore and aft, and joined the parade of boats heading for open water. We spent the next hours about three miles offshore, where Antonio told me that there was a "platform" and only about twenty feet of water; it was here that octopuses liked to hunt. Antonio fished seven lines off two jimbas. The lines were weighted and each had a crab tied tightly on the end. He hurled them about ten feet away from the boat and moved from one to another, testing each with a forefinger. On one side of the boat, he tied off a six-foot-wide canvas weighted with sandbags, which served as a drag against the current and kept the boat steady. Even though the weather was deemed decent for fishing, the swells were substantial, and the sandbags helped to keep the boat level. The movement of the boat in the water also meant that the crab on the ocean floor moved and attracted octopuses.

When a tug on a line indicated that an octopus had enfolded a crab, Antonio had to judge the right moment to pull up the line without dislodging the octopus. It had to be when the crab was firmly in the octopus's grasp but before it had eaten too much. An octopus has a methodical way of capturing and eating a meal, and jimba fishers take it into account. The way an octopus finds food is what researchers have labeled "speculative foraging." This means they go out looking, and some days, of course, it will take longer than others to find something. Not usually too long, though, because octopuses are superlative hunters. They sense prey by using a combination of three senses: by sight, or by touch with a long arm extended and searching, or by the equivalent of scent, a chemical stimulation of receptor cells on the suckers of their arms. Experiments have shown that even when an octopus in a tank cannot see prey on the other side of a divider, it senses a crab's presence there immediately.

When an octopus in the wild sights or senses a potential meal, the idea is to get the prey under its mantle cavity as soon as possible, where it will be attacked by the beak and mouth. The mantle itself is the sac holding its digestive organs. The muscular mantle also contracts to draw water into the gills, where oxygen is extracted for the blood. Two of an octopus's three hearts are used to pump blood through the gills, while the other pumps it through the body. When the water has been filtered through the gills, it is expelled by the siphon. When done with force, this expulsion of water also serves to move an octopus along rapidly.

Sometimes, when it is spots a prey, it will pounce directly, using what marine biologist Roger Hanlon called the "parachute attack": descending on, enfolding, and covering its victim.[8] Henry Lee, a naturalist and director of the Brighton Aquarium in England, described it in 1875: "The action of an octopus when seizing its prey for its necessary food is very like that of a cat pouncing on a mouse, and holding it down beneath its paws. The movement is as sudden, the scuffle as brief, and the escape of the prisoner even less probable."[9]

Other times, an octopus will lie on the bottom, looking for all the world like just another rock, and suddenly an arm will whip out to capture a passing crab and stuff it under the mantle. Once beneath that formidable web of flesh, torn at by beak and suffering the effects of octopus venom, there is no escape. Octopuses have small mouths, and it takes them a bit to consume a crab, as they can only swallow small pieces of meat. The esophagus of an octopus is narrow and passes directly through the brain on its way to the gut. Octopus saliva is a strong digestive agent, breaking down flesh and making it easier to swallow. Overall, digestion is a slow process for an octopus. So that it does not have to pass up a meal should one cross its path, the octopus gut has a crop where food can be stored for later use.

"I know just how much these lines weigh, how they feel with a crab on there, and if an octopus has hold of the crab, I can feel it right way," Antonio told me. In fact, for the entire three hours we were out, he didn't pull in an empty line; every time he pulled one in an octopus was wrapped around the crab at the end of it.

Antonio learned how to fish for octopus from his father, who was one of only a few doing it during his day. Octopuses were plentiful, but there was not much market for them, except locally, and they were cheap. Still, they were there for the taking. Those his father couldn't sell went straight to the family's dinner table. The boats were originally wooden, propelled with a sail, and the lines were made of henequen and the poles of mangrove.[10] Even with the modern improvements of fiberglass boats and outboard motors, it was hard work, as I saw for myself.

The sun beat down on the water and the boat did not have an awning or shade of any sort. Nor did it carry any life jackets, even though we could barely make out the coastline as the boat rose and fell. In case the boat sank, swimming to shore would not be an option, even if a person could swim. Six Yucatecan octopus fishers had met their deaths at sea in 2022. Antonio admitted that he and many of his fellow octopus fishers could not swim. "There's a law that says you have to have life jackets on board, but the truth is we're an unruly, stubborn bunch, and we don't mess with them."

When he pulled in a line, Antonio removed the octopus from the crab with one hand and inspected its third arm, running a thumb down it to see if it was the grooved hectocotylus of a male, before tossing it into a large blue bait barrel, which he had filled with seawater, occasionally pouring more pails of water into it to keep the water oxygenated. Today, the octopuses he caught would go to the Octopus Project, and what they particularly wanted were females.

Mayan octopuses lay their eggs in December, so if the cooperative wanted to be raising juveniles in the spring, they would need to have eggs in December. But females were proving scarce, and of the dozen octopuses Antonio caught that morning, only two were females. Regardless of gender, they would all be taken back to the research station.

· · · · · · · ·

It turned out that the extraordinary evening sunsets I remembered enjoying in solitude from the beach in 2010 were a thing of the past. The sunsets were as amazing as ever when I returned to Sisal in the fall of 2023, but the solitude was gone. One of the first things I did when I got back was to visit Don Baby and Doña Genny in their small house, and she warned me that things had changed. Little by little, Sisal was gaining ground as a tourist attraction, and there were always a sizeable number of outsiders in town. Among the things that visitors liked to do best, and which had become virtually obligatory for tourists, was to sit on the beach or the town pier and watch the spectacular sunsets. Crowds of loud young people gathered each night to cheer as the last sliver of sun sank below the gulf's horizon.

"It was better when you were here before," Doña Genny told me, when I reminded her how she had improved my daily life by encouraging me to walk to work along the beach. "Now, there are a lot more tourists all the time. Things are different. There are lots of strangers in town. I see people in the street and I don't even know who they are. That never used to happen. I lock the door when I'm here alone. I believe in spirits, but the dead don't scare me. It's the living who frighten me."

Bicycles were no longer left outside unlocked. A supermarket had opened up close to the center of town. It had an ATM machine, and you could pay for your groceries with a credit card, something unheard of when I lived there in 2010. The beach, too, had changed. What was once scrubland and dunes between the beach and the dirt road that led out to the UNAM station now held houses, and they weren't small. Enough of a tourist market had grown up so that people could invest in building a big house in order to rent or sell it to middle- and upper-class families from Mérida for use as a second home. The beach was now full of *palapas*—thatch-roofed open shelters built around thick wooden posts permanently sunk into a concrete base—rented out to beachgoers. In 2023, the beach was rarely empty, but at least those stretches farther away from the center of town were not crowded. On the disputed land that had originally been planned to hold an expanded octopus farm, modest cement-block houses and fancier two-story places had been built, mostly belonging to non-Sisaleños. What I paid to rent a room in the center of town in November 2023, a couple of blocks from the beach, was ten times what I had paid in 2010.

Like so many other fishing villages around the globe with a big city close at hand, tourism was growing, generating more revenue every year. A market had developed to attract the growing middle class from Mérida who were eager to have a place to breathe sea air, somewhere they could find a refuge from noise and traffic. Mérida was a big, vibrant, beautiful city with over a million people, but it had way too many cars, old exhaust-spewing buses, and limited garbage collection. In 2019, the federal government's tourism department announced that Sisal was being designated a "Magical Village," one of only 103 in all of Mexico.

Far from celebrating the designation, Sisaleños began organizing a campaign to reject the title of *Pueblo Mágico*. Banners announcing the honor suddenly disappeared, and plaques commemorating the designation were covered up. In the summer of 2021, the road to Hunucmá was blocked by a group of angry Sisaleños who claimed that the title would lead to the destruction of the mangrove ciénega and the privatization of more land for construction, and that the economic benefits would go only to those who were already prospering. The resistance to the designation was led by the village commissioner, Miguel Antonio Ek Peck.

"In December 2019 we learned we were certified as a Magic Village," he told me. "We initially hoped this would bring some federal money for things like safe drinking water, garbage collection, and health care. But eventually we understood that the certification was just to benefit private interests and increase the price of things like land being sold off, which is something that was happening anyway, and is still happening.

"One thing it does is bring in people who we don't know. Sisal has been a very tranquil place in all its aspects, but today what you see is people coming here who we don't know and who bring problems in some form into the community. Could a Pueblo Mágico designation help us? The reality is no. Sisal already attracts enough tourists.

"We have a long and magical history, from being an important port, through henequen. We don't need to be certified, which is something that will only be a blow to the inhabitants and contribute to eventually displacing them."

What Ek would like to have happen, instead, is the creation of a development plan for the village, which would call for things that would benefit all Sisaleños. But, by law, he cannot even draw up such a plan, much less dedicate the village's resources toward implementing it. The plan and its administration, like all administrative decisions about Sisal, has to be managed from Hunucmá, nearly fifteen miles away. According to Ek, there is little political will there to make such changes. Sisal, as a community, is tied hands and feet because it is administered by Hunucmá, which has plenty of its own problems to consider. Miguel Ek, frustrated and stymied by the system that makes it impossible for Sisal to advance on its own, told me he had decided not to run for office in the future.

One of the brighter hopes Miguel Ek had for Sisal was the Octopus Project. More and more octopus is being eaten around the world every year. While still a rarity between the coasts of the US, and not often consumed in underdeveloped countries away from the coasts, it is poised to make the jump to general consumer acceptance. As it is, some twenty-five thousand tons of octopus from the Yucatán Peninsula are sold annually around the world, worth about $55 million. The octopus fishery in the Yucatán has been estimated to benefit directly some twelve thousand fishers and their families, along with making indirect economic contributions to other economic sectors. Mexico ranked third in the world for octopus sales, and around 95 percent of that came from the Yucatán.[11] Sisaleños, and a multitude of people living in fishing villages up and down the Yucatán coast, could have their lives changed for the better through small-scale octopus farming.

"With the traditional fishery here it is difficult to earn a living," Miguel Ek told me. "These are bad times for many who used to work fishing and who have had to come ashore and work with tourists. [The octopus farm] is very important. Now, the Moluscos del Mayab have their own place to farm; they have their land close to the sea. Hopefully it will grow and be a sustainable and profitable source of revenue for our village. The cooperative has worked long years developing the ways to farm octopuses, and we can only hope all their hard work will be rewarded."

GALICIA, SPAIN

G ALICIA has always been octopus country.

With seven hundred miles of Atlantic Ocean coastline and eighty ports, octopuses are an integral part of the Galician diet, identity, and economy, and have been so for millennia. Galicia is all the way across Spain from Barcelona where I have lived for over thirty years. The autonomous region of Catalonia, of which Barcelona is the capital, also has an octopus fishery, but the relatively tranquil, sun-blessed waters of their Mediterranean fishing grounds are a far cry from the thunderous, tumultuous grays of Galicia's storm-tossed Atlantic. People who know their octopus will tell you that those from the Atlantic taste better. Not that the Mediterranean animal tastes bad, they'll hasten to add, but it just doesn't have the rich flavor of those from the Atlantic.

The two autonomous regions on opposite sides of Spain are different in many other things—each has its own markedly distinct flora, fauna, topography, regional character, and language. Fortunately for me, Spanish is spoken all over the country, and I have no trouble communicating. While octopus is avidly consumed on both coasts, even in Barcelona people prefer the Atlantic animal. *Pulpo* is Spanish for "octopus," and the Pulpería Bar Celta is a tapas bar specializing in octopus, tucked away on a medieval Barcelona back street near the port. Its signature dish is octopus, serving over ten thousand pounds of it a year brought from the Atlantic Ocean, all the way across Spain. Luis Gallego, the head cook and octopus buyer for the pulpería, said that his preference for Galician octopus had nothing to do with his last name, which means Galician. "I just don't buy octopuses from the Mediterranean. I'm not interested in them," Luis told me, having spent seventeen years preparing octopus in the traditional Galician style at the Bar Celta. "It has to come from the Atlantic."

Galicia is often referred to as the seventh Celtic nation, the others being Ireland, Wales, Scotland, Cornwall, the Isle of Man, and Brittany in France. In fact, genetic evidence seems to suggest that people from northwestern Spain—the Basque country and Galicia—were among the very early settlers of Ireland, perhaps around 1000 BC. An Irish myth relates that it was the sons of a Galician king, Milesius, who sailed to and settled Ireland.[1]

The Celtic Galicians lived in hilltop settlements called *castros*, not far from the ocean, which they relied on to provide sustenance. In the remains of one small castro, from the fifth century BC, some sixty postholes were found, which were believed possibly to have been for posts holding racks of drying octopuses.[2] Even in 2023, driving through small Galician villages close to the coast I occasionally passed a typical low, stone house with half a dozen octopuses drying on a wooden rack outside, as if they were clothes hung out to dry. If an octopus is to be consumed within a few days, forty-eight hours drying on a rack will be enough, but to be thoroughly dried, six days are necessary. It is well dried when you can pick it up by an arm that doesn't bend. A thousand years ago, travelers passing through Galician coastal villages would also have seen stone houses with octopuses drying in the yard, likely a much more common sight then.

The Roman Empire began spreading its influence across the length and breadth of Spain, expanding into Galicia in the first century BC, bringing and imposing a different way of life. The castros gradually disappeared. Romans, like the Greeks before them, were fond of fish and seafood. Not only did the Greeks like to eat octopus, but they considered it an aphrodisiac, according to the food historian Andrew Dalby who wrote that Athenaeus cited Alexis, a comic playwright from the fourth century BC reputed to be an epicure: "Octopus increases sexual vigor, but it is tough and indigestible. The larger species provide better nourishment. When boiled slowly it settles the stomach and moistens the bowel."[3]

While the Greeks had to content themselves with octopus from the Aegean, the Romans found a huge population of delicious octopuses off the coast of their new colony in Galicia, and it must have pleased them mightily. They held octopus in high culinary regard, often preparing it in a big pie, cutting up the arms and filling the head with spices. Pliny the Elder, who died in 79 AD, noted that the Romans used bamboo knives to slice an octopus because they believed metal left behind a taste. The Roman colonization of Galicia may be the first time that Galicians began to look at the octopus as a commodity instead of simply supper, as something that could be sold instead of eaten. Octopus was no longer just something to put on the table; it had significant commercial value. That valuation has held over the past two thousand years.

The worth of octopus to the Galician economy grew over the centuries. By the Middle Ages, it was so highly valued inland, far from the Galician coast, that it served as currency. There are records from as long ago as the last half of the fifteenth century showing that sharecroppers paid with dried octopuses to rent the land they farmed from the monasteries that owned it. Octopus was particularly valued by the ecclesiastical class during the Lenten period when land-based meat was prohibited. Muleteers, called *arrieros*, traveled inland with carts full of dried or smoked octopus and other seafood, for which they found a ready market, and they returned to the coast bringing the produce of the inland farms, things like hams, sausages, cheese, potatoes, and wine. As the Galician economy grew, so did the prominence of octopus. In Marín—still today an important Galician fishing port—some forty boats captured more than two hundred thousand pounds of octopus in 1787, which was consumed locally and also exported, according to records from the Oseira monastery.[4]

Galicians are often stereotyped by other Spaniards as a dour, introverted people, pessimistic and gloomy, gray as the weather in which they live. Galicians have some good reasons for it, if it is true. Their natural resources have often poured profits into the pockets of businessmen from other regions, while Galicians work to generate those profits under hard conditions for little pay. In the nineteenth and twentieth centuries, more Galicians emigrated to the Americas looking for work than from any other part of Spain. Fidel Castro's father was Galician, among many who settled in Cuba in the nineteenth century. Work has always been scarce in Galicia; farming its rocky ground battered by tempestuous weather has never been easy. Fishing holds its own multiple perils, and industry is relatively scarce.

Aquaculture was one thing Galicians were well positioned to do in the late twentieth century without leaving home. With a strong marine biology department at the university in Vigo, Galicia's largest city, all that coastline, and an abundance of wild fish and seafood in its waters, Galicia naturally became an early center for Spanish aquaculture. Farmed mussels, turbot, sea bream, and trout are grown there in large numbers, and by the beginning of the twenty-first century they had all carved out a respectable share of their markets.

A big octopus market was waiting. Spain was behind only Japan in the amount of octopus consumed annually—importing some fifty-five thousand metric tons in 2016.[5] Galician waters held plenty of octopuses, and there is an ancient body of knowledge about how to catch, process, and market them. At first, researchers were confident that it would be relatively easy to farm them. By the mid-1990s, private investors, together with the university, were funding efforts to grow octopuses off the coast of Vigo in outdoor cages, which were stacked in coastal waters. It proved much more difficult than anticipated.

"To create the right conditions for them costs a lot of money and a lot of effort," Ángel Guerra, a marine biologist who was involved in those early attempts, told me over the phone. "In the future, if octopus reaches a very high price on the market, it may yet be economically viable. When we started, we bought little ones from fishermen, separated the sexes and put them in cages in the ocean. We fed them with bycatch from the fishermen, and in six months we could raise an octopus for the market. What we needed was to cultivate them, produce eggs, and close the cycle. The question became increasingly complicated and costly."

Eventually, the outdoor cages were abandoned, and researchers from the state-owned Spanish Institute of Oceanography began working to produce viable results in indoor tanks. For a long time, things did not go much better. It took years to learn how to keep the paralarvae alive, and even when this was accomplished, survival rates were abysmally low. In 2007, a group of Spain's leading octopus researchers published a paper to explain why octopus aquaculture was proving so difficult. They admitted, "Despite the research effort taken until now, the rearing of *O. vulgaris* paralarvae is generally characterised by high mortality rates and poor growth, which hamper the entire culture of this species."[6]

Finally, in 2017, researchers at the institute announced that they had succeeded in consistently raising paralarvae into reproductive, adult octopuses, closing the life cycle. Potential investors sat up and took notice. In 2018, the publicly owned institute signed a fifty-year exclusive agreement assigning its patents for octopus aquaculture to the privately owned Nueva Pescanova,

Spain's largest seafood company, with annual global sales of frozen fish and seafood of over a billion dollars. Headquartered in Vigo, the company was already heavily invested in the aquaculture of various species. Nueva Pescanova constructed a marine research center just outside the coastal town of O Grove, near the Galician city of Pontevedra, to be principally dedicated to octopus research. The center was inaugurated in the summer of 2021 with great fanfare, at a ceremony attended by the president of Galicia. A marine biologist and veteran administrator at Nueva Pescanova, David Chavarrías, was named director of the Pescanova Biomarine Center, and it was staffed with researchers, including some hired away from the Oceanographic Institute.

Then, in early 2022, Nueva Pescanova made an astonishing announcement. The company was going to invest €65 million ($70+ million) in building the world's first industrial octopus farm and processing plant, beside the port of Las Palmas on Spain's Grand Canary Island.[7] The plant was projected to occupy thirty-seven thousand square yards and take two and a half years to construct.[8] The company confidently predicted that by 2025, it would be growing octopuses in a thousand tanks and putting a million pounds of octopuses on the giant global market every year thereafter. The farm was only waiting for the final environmental permits to be approved by the Grand Canary authorities.

The announced marketing plan was radically different from that of Sisal's Moluscos del Mayab cooperative. Carlos Rosas had given a lot of thought to the octopus market he wanted to reach in Mexico. His aspirations involved creating a new source of income for the members of the Moluscos del Mayab cooperative and for the community of Sisal. They did not include going worldwide with millions of pounds of octopus. "We want to produce octopuses that don't enter the huge global market and won't require a massive aquaculture," he told me. "Rather, we're aiming for a market that's a little more elite and can afford to pay a little extra. We want to have a stable business for the fishermen and their wives but without the risk of massification. Here [in Sisal], for more than half the year, out of season, no wild octopus is fished. Aquaculture has an advantage—it permits a standardized production all year. And, because it's done on land, it allows women to participate, a part of society that normally can't collaborate in these sorts of things.

"We're environmentally conscious, more so than a factory farm is likely to be. We're recirculating the water and producing very little waste. Water recirculation systems that were expensive a decade ago are now cheap. We're also using some technologies that are expensive on the market but that we make here for lots less. They may not be as efficient as something you'd buy on the market, but they're made with recycled materials at low cost, and this has enabled us to reduce our expenses by a lot."

Another difference between the aquaculture projects in Sisal and Galicia was that while Carlos Rosas had invited me to come and nose around for as long as I wanted, and even work there, Nueva Pescanova was extremely secretive. After trumpeting its plans for industrial octopus farming in the 2022 announcement, the company faced an immediate storm of protest from animal rights advocates around the world. Nueva Pescanova quickly became reluctant to allow any journalists on the O Grove property or to concede any interviews. Nevertheless, when I wrote an email from Barcelona to the center's director, David Chavarrías, and told him that I had spent time with Carlos Rosas in Sisal, he gave me his okay to come to Galicia and see for myself what they were doing.

That's how I found myself in the spring of 2023 standing in the basement of Nueva Pescanova's marine research center in O Grove, gazing into large tanks at the fifth generation of aquacultured octopuses hatched and grown at the center. These octopod lodgers at O Grove were the brood stock for what would be Europe's only octopus farm, as soon as the company received the necessary environmental permits and got the farm up and running. Plans called for thirteen acres by a pier in Las Palmas, with as many as a thousand tanks and a large processing plant. The octopuses living in tanks in the basement at O Grove would be the farm's foundational animals, the original progenitors of millions of farm-grown octopuses. The light in the center's basement was kept dim, as octopuses like it, and when I visited there was a low, calming background sound of circulating water.

In the wild, the octopus is a short-lived animal, notoriously antisocial in most cases, preferring to spend its year or two of life in the solitary splendor of its den and coming out at night to hunt. Octopuses are homebodies, the only cephalopods that make a dwelling. They generally stay inside during daylight hours, devoting a lot of time to cleaning their shelters and often bringing empty shells and detritus to put around the outside of their dens to mark its borders. Jennifer Mather, a Canadian marine biologist and behavioral psychologist, found that both male and female octopuses spend 3 percent of every day's waking hours on housekeeping. They will use their funnels to blow rocks, empty shells, and waste out of their dens until a midden forms outside.[9]

Scuba divers use these middens as a way to locate octopus dens, and octopus fishers in some places take full advantage of the octopus's domestic nature. In Catalonia, for instance, those who fish for octopus in the Mediterranean do not even use bait; they simply lower a line of the same ceramic half cylinders used for roof tiles all over Spain—an attractive piece of real estate in the eyes of an octopus—to the ocean floor. Fishing with some form of octopus pot is practiced by many artisanal fishers in a number of places around the world. Because they require no bait and can be used repeatedly, octopus pots have a lower carbon footprint than traps, which need bait, or trawlers, with their high fuel consumption and on-board freezer facilities.

When an octopus sees an empty pot, it sees a new home, and once installed it will not abandon its shelter even as it is hauled to the surface and brought aboard a boat. The Egyptians, and later the Greeks and the Romans, used this same method with ceramic amphorae tied to a rope and lowered to the seabed, as attractive a home to an octopus thousands of years ago as it still is today. In Tunis, the Greek Orthodox faithful were particularly devoted to eating octopus during Lent, and they also fished with baitless ceramic cylinders. In Japan, most fishers have switched to plastic pots, but a few octopus fishers still use *takotsubo*, long, narrow earthenware pots tied to a rope and lowered to the bottom. Later, when octopuses are comfortably settled in, they haul the rope on board. Today in Japan, takotsubo is also the name of a heart disease in which the left ventricle balloons out in the shape of an octopus pot.

Despite their natural predilection for domesticity and solitude, the octopuses in O Grove seemed to have had no problem adjusting to communal living in Nueva Pescanova's tanks. I saw half a dozen or more unusually large octopuses in each tank of seawater at the center. Some swam languidly about, while others were grouped peacefully together in a sloppy pile at the bottom of the tanks. Octopuses can squeeze their substantial bulk down to almost nothing, and they are justly famous for being able to escape from virtually any enclosure. The water came up almost to the top of these tanks, and it would have been a simple enough matter for an octopus to sling one of its long arms over the side one night and pull itself out. It had yet to happen at Nueva Pescanova, and that was because the animals were content, according to Pablo García Fernández, who was one of those veteran researchers hired away from the public Oceanographic Institute to work at the O Grove facility.

"At the public research institute where we were working there were just two of us working with octopuses, and we had very few resources, not like here," he told me as we gazed into a tank. "We had reached the stage where [paralarvae] went to the bottom, but we hadn't completed the cycle. We were just a couple of guys with nothing but our salaries."

When Nueva Pescanova began paying attention to the potential of octopus aquaculture, research positions were offered to Pablo García and his colleague. "When Nueva Pescanova got interested in the project things started moving faster, and the next year we had more to work with. Our goal changed from just seeing how we could keep a larva alive for thirty days to really being able to complete the cycle."

When the researchers found themselves with more ample resources, progress was rapid. The big octopuses I was watching in those basement tanks had never known the ocean; they were born and bred in tanks, García told me as we watched them swim lazily about. "These octopuses have everything controlled. The light and the oxygenation. They have all they want to eat. The water temperature is perfect. Apart from the fact that they'll be eaten in the future, they couldn't have better conditions," he told me, as we stood beside a tank containing several unusually large octopuses, each weighing more than twenty pounds.

Those impressive specimens may indeed have been living the good life in the research center's basement, but protest after protest followed the company's announcement of its plans to farm octopuses on an industrial scale. One hundred international scientists and scholars signed a widely published open letter opposing octopus farming for both ethical and environmental reasons. Hundreds of thousands of signatures were garnered on petitions calling for a prohibition of octopus aquaculture. Why, asked environmentalists and animal protection groups around the globe, should Nueva Pescanova grow, slaughter, and sell a solitary animal with the highest intelligence of any invertebrate on the planet?

The answer, of course, was that the market price of octopus had soared over the past two decades. The animal's presence is obligatory on the menus of upscale restaurants and sushi bars in the United States and Europe. Various European Union countries exported some $91.5 million worth of common octopus to the US in 2017, and between 2012 and 2018, the volume of Spain's octopus exports to the States increased by 227 percent. Prices during these years were climbing steadily, but European catches of octopuses during those years declined by more than 40 percent.[10]

The annual global market for octopuses had more than doubled since the 1980s when it absorbed 170,049 metric tons, reaching 405,773 metric tons by 2019, and between 2010 and 2019 the value of the global octopus trade had doubled to $2.72 billion from $1.30 billion. By 2029, annual sales are predicted to exceed $12 billion.[11] All these octopuses will be caught in the wild, and whoever can operate a successful farm stands to make a bundle of money.

By 2014, executives at Nueva Pescanova began to take the potential of octopus aquaculture seriously and to lay the groundwork for an octopus farm. "That's when I began to have a sense that it could function," the marine center's director, David Chavarrías, told me in 2023. "Things began to work and to come together. The cycle still wasn't closed, but we began to see the possibility of success, and that's when we began to really get interested. It's a species that grows rapidly and is very attractive for aquaculture, so I got in touch with the institute's researchers like Pablo and everything began to come together."

David Chavarrías was fifty years old when we spoke, originally from Madrid, and he had been working for Pescanova since shortly after finishing his dissertation on fish genetics. When the company had offered him a job in Galicia researching the genetics of turbot, with an eye to cultivating them, he jumped at the opportunity. "I didn't have to think twice about it," he told me. "Just imagine what it means for a young biologist to get to work in his field with the support of a multinational."

The turbot is a flat fish, which Pescanova's fishing vessels caught in large numbers. By 2019, Nueva Pescanova's affiliate, Insuiña, was farming over six million pounds of them a year. "I was the head of the team that began to work on genetically improving turbot. I worked on turbot genetics for ten years, and in 2008 I was promoted to the director of the plant. I went from research to administration. Then, in 2018, with the turbot plant established and doing well, the company began to develop a plan for octopus.

"At that point we still had a long road ahead. Really [with octopus] you're dealing with three totally different phases, three different animals. There's the paralarvae, the larvae, and the benthic, and to cultivate them you have to learn about how to treat each phase so that you obtain high survival rates. The company began to apply for patents and things began to fall into place. Our first successful culture was a female who we named Lourdes, because we were hoping to do something miraculous with her, and it was with her that we closed the cycle for the first time. From there, we carried on researching how to create optimum conditions for the well-being of the animal, how best to raise it and to sacrifice it, and now in 2023–24 we've accomplished this work. Now we're ready to take it to an industrial scale."

That was, perhaps, good news for the multitude of people who love to eat octopus, but in addition to consumers who extolled its taste, many people professed to love the live animal. They read of Nueva Pescanova's plans with dismay and anger. Attitudes toward octopuses and their images have gone through a 180-degree cycle. Most written references throughout history have praised the octopus's flavor and warned of its ferocity and danger to humans.

As early as the thirteenth century, and typical of the prevailing opinion, the Flemish monk and theologian Thomas of Cantimpré wrote about the octopus in his natural history: "This animal has such strength in its arms that it sometimes seizes by force a sailor standing carelessly on a ship and drags him into the sea, and is satisfied with his flesh: for it eats meat gladly."[12]

In his 1788 catalogue of Galician marine species, José Cornide, a Galician geologist and naturalist, wrote of the octopus: "There are some that are huge and when they get a grip with a free arm on a man, it [the arm] creeps along his body until it gets close to his throat, or it manages to submerge him in water until he drowns."[13]

In the nineteenth century, popular authors continued to represent the octopus as a fierce, threatening creature, an enemy and predator of human beings. Known as the devilfish, it was represented as evil incarnate. More than any other single work, what really spread the notion far and wide of the danger to humans represented by octopuses was Victor Hugo's bestselling novel *The Toilers of the Sea*, published in the original French in 1866 and translated that same year into English. The protagonist engages in a terrible battle with an octopus, an animal which Victor Hugo portrayed as evil incarnate, a notion driven home by Gustav Doré's horrifying illustrations, which accompanied the original text. Victor Hugo wrote, "They are the chosen forms of evil. . . . They are as the darkness converted into beasts. . . . These animals are phantoms as much as monsters. . . . They mark the transition of our reality to another."[14]

For centuries, this was how the Western world had represented octopuses—an animal to be feared and avoided in the water, a dark being from another world with no relation to our own, wholly foreign to human beings. For that reason alone, many people were terrified by octopuses, and their first reaction to something so different, so unknown and unfathomable, was fear. Throughout the Middle Ages, this did not stop Galicians from continuing to value octopus as a food, and it appeared regularly in markets. In 1591, for instance, in Vigo, it figured as one of the most frequently caught species, along with sardines.[15] However, any appreciation for the creature extended no farther than a dinner plate.

Over the second half of the twentieth century a lot of research was done on both octopus body and octopus brain, and this has continued into the present. The more biologists studied the animal, the more remarkable were the qualities revealed, including the octopus's capacity for learning. Many researchers concluded that it had a high level of cognition, equal to that of a dog or a cat, unique among invertebrates, and some came to believe that, in addition to brains, octopuses have minds. During the period from 2013 to 2020, about a hundred journal articles were published dealing with *O. vulgaris*.[16]

The octopus is descended from animals that had shells, and their strategy for staying alive in those days was defensive, since they were well protected. Without shells they had to develop new strategies for survival. One of those was the amazing speed with which they could blend into their surroundings. One set of researchers wrote, "Their skin is a perfect tool for anti-predatory strategy as it provides an ideal means for camouflage, to make cephalopods difficult to detect in their surroundings, or, in opposite, become very bright, to make predators think they are venomous."[17]

To make up for their missing shells, octopuses also had to modify and improve their sensory and nervous systems. They had to rely more heavily on their vision and generally had to get smarter, developing a larger brain with a greater capacity for learning and adaptation. Just as octopuses evolved a capacity for cognition in response to being naked, so did humans. Without the ability to learn and adapt our behavior, we would long ago have become extinct.

Perhaps it is this part of the evolutionary path of both species that has finally converted an animal we once feared into one that elicits responses of empathy and respect from our species to theirs. In the digital age, the octopus has become a positive symbol of the complex and marvelous natural world of which both humans and octopuses are a part. In 2020, Netflix aired a blockbuster documentary, *My Octopus Teacher*, and a number of other sympathetic books and documentaries have purported to show the octopus mind at work.

One online group named OctoNation, billing itself as the world's largest octopus fan club, declares on its website, "Our nonprofit mission to inspire wonder of the ocean by teaching the world about octopuses. . . . Now with over 600,000 members and content that reaches millions weekly, we collaborate with underwater photographers, artists, marine biologists, aquarists, and organizations who are inspired by the octopus and our mission."[18]

The idea of raising them at a factory farm in a confined space—and slaughtering them to sell in a supermarket for a family's supper—horrifies many people. All hell broke loose following Nueva Pescanova's announcement of its intentions. An open letter signed by a group of environmental scientists, which was partially reprinted in *Business Insider*, deplored the idea of a farm for both environmental and ethical reasons. It said in part, "Beyond environmental and health concerns, octopuses are capable of observational learning, have individual personalities, play, and are capable of problem-solving, deception, and interspecies hunting."[19]

One prominent spokesperson for the antifarming campaigners was Jennifer Jacquet, an associate professor of environmental studies at New York University. "A life in solitary confinement for a curious mind is ethically wrong," she told *Time* magazine. "Beyond their basic biological health and safety, octopuses are likely to want high levels of cognitive stimulation, as well as opportunities to explore, manipulate, and control their environment. Intensive farm systems are inevitably hostile to these attributes."[20]

Nueva Pescanova fought back. In a statement to *Business Insider*, from its headquarters in Vigo, the seafood company responded to the animal rights critics. Nueva Pescanova stressed that it had a commitment to humane farming practices and challenged claims that octopuses are particularly smart, stating that while they are demonstrably capable of adapting to their surroundings, "there is no scientifically validated knowledge about the 'intelligence' of the octopus, or whether it is more or less intelligent than other marine species that are already farmed."[21]

The amazing ability of the octopus to learn was already well documented in the last half of the twentieth century. Many experiments were carried out by prestigious researchers to map the octopus brain and gauge its capacity for learning. Much of this work was carried out at the Stazione Anton Dohrn, a marine research center in Naples, Italy. The Stazione was founded in 1872 by Anton Dohrn, a nineteenth-century biologist who had the support of Charles Darwin. The Stazione is an island of calm and tranquility located in a small park close to the sea in the heart of Naples. Outside the park is the high-octane pace of Neapolitan life, and in the last half of the twentieth century it drew some of Europe's top octopus researchers, often professors from other countries, who chose to spend their summers doing research at the Stazione and enjoying the pleasures of Naples. From 1901 to 1991, twenty researchers who had spent time at the Stazione were awarded Nobel prizes.

Typical of the work done was that carried out by the British zoologist Martin J. Wells, grandson of the writer, H. G. Wells. Martin Wells demonstrated that octopuses could discriminate by touch between grooved and smooth surfaces and, more remarkably, that they could remember and apply what they had learned. He noted the comparable evolutionary paths taken by humans and octopuses:

There is a haunting similarity in the history of *Octopus* and ourselves. Both of us have evolved from groups obliged to spend a period in the wilderness, eking out a peripheral existence while other animals, temporarily better adapted, dominated the more desirable habitats. The teleost fish and the great reptiles respectively forced upon us a way of life that depended upon adaptability rather than armor, a capacity to know when to run and how to detect trouble in the making that has eventually placed the coleoids and the mammal in a unique position with respect to their invertebrate and vertebrate competitors.[22]

Other experiments at the Stazione by the eminent British zoologists J. Z. Young and B. B. Boycott demonstrated the ability of octopuses to find their way through mazes to reach a crab. Young wrote enthusiastically, "The brain of the octopus has already abundantly proved its value for the study of behaviour. It is perhaps the type most divergent from that of mammals that is really suitable for study of the learning process."[23]

One of the corollary discoveries of these experiments was that different octopuses approached and accomplished a task differently. It began to be posited that octopuses had distinct personalities and learning capacities. Perhaps the most controversial of experiments at the Stazione were those carried out in 1992 by the Italian neurobiologist Graziano Fiorito, who concluded in his experiments that some octopuses have an ability to learn by simply watching other octopuses perform a task. To be capable of observational learning meant using a sophisticated mental process far beyond what had previously been ascribed to octopuses.

Graziano Fiorito had impeccable credentials. Born and raised in Naples, he turned down an offer to join the eminent biologist E. O. Wilson's team at Harvard and chose to work at the Stazione. He was originally working with crabs, but in 1985 he saw a picture of an octopus opening a jar to extract a crab. "That was tremendously interesting," he told me when I visited the Stazione in 2010. "It's important, because this kind of problem-solving in terms of behavior is highly complex and demonstrates impressive cognitive capabilities. In 1986, I finished my last experiment with crabs, and thereafter I have dedicated all my life to octopus."[24]

In Fiorito's experiment, octopuses learned to choose between two colored balls, one of which rewarded the animal with a fish, the other of which delivered a painful shock. It took an octopus an average of sixteen times before *always* choosing the ball that brought a reward. Next door, and able to observe what was going on through clear plexiglass, was another set of octopuses, and Fiorito reported in an article published in *Science* that the observers did much better when they were put through the trial than the first group, concluding that they had learned by watching.[25] His paper immediately drew strong criticism from many researchers who could not believe that an octopus was capable of observational learning, but he stood by his results and told me in 2010 that he had recently shown that octopuses can learn simply by watching a videotape.

Fiorito's 1992 results were disputed in many quarters, with his critics questioning the experiment's methodology, its absence of adequate control experiments, and the possibility that researchers were visible to the octopuses during the trials and might have inadvertently given them visual cues. Be that as it may, Fiorito's work certainly made other researchers reassess octopuses and their cognitive abilities.

An octopus has a central brain and eight primitive brains in its other arms, which are packed with about forty million neuron cells each. "The study of the octopus cognition structure shows that all components of the octopus body are connected with each other," wrote Oxana Shamilyan in 2021. "Neuron signals are transmitted from brain to arms to implement some actions, while others are initiated in the arms themselves, from arms to brain to process sensory information, from eyes to skin to change appearance and so on. The whole body works as a single mechanism and implements various complex tasks."[26]

The Canadian octopus researcher, Jennifer Mather, teamed up with Roland Anderson, director of the Seattle Aquarium, and in 2008 they added their own startling claims to the discussion. They designed an experiment that confirmed the suspicions of earlier researchers that octopuses had different ways of reacting to stimulus. They elaborated three forms of behavior in which octopuses differed one from another, which they named "active," "reactive," and "avoidance," and used the behaviors to elaborate distinct personalities. For instance, when a small wire brush was inserted into a tank, octopuses exhibited a range of reactions from crawling away, to an ink blast, to aggressively grabbing the brush. When a crab was introduced into a tank, some octopuses would jet through the water to capture it, while others walked along the tank's bottom to reach it, and still others waited until the crab came within grabbing distance.

In 1998, Mather and Anderson designed an experiment with giant Pacific octopuses to determine if they were capable of play. They put an empty plastic pill bottle in a tank with a slow current that carried the bottle toward an octopus. Two of the eight test subjects jetted water at the bottle to blow it across the tank, then waited until it drifted back to them before jetting water at it again. Each of the two repeated the action more than twenty times, Mather and Anderson wrote in the *Journal of Comparative Psychology*.[27]

Whether an octopus's behavior actually displays intelligence still depends on whom you ask. The debate revolves, in part, around the thorny question of what constitutes intelligence. Further, one must ask whether an octopus's behavior is not really intelligence but adaptive learning (no small feat in itself) to enhance survival.

When an octopus teaches itself to unscrew the cap on a jar so it can reach the crab inside—a sequence viewed many, many times on the internet—is that intelligence, or is it simply adaptive learning?[28] And is there much difference between the two? Is it fair to compare octopuses to dogs, which form meaningful social networks with other dogs and have discernible emotional relationships with humans? How many of the two-way relationships between octopuses and humans that have been portrayed in contemporary books and documentary films are only one-way in reality: the human projecting on to a cephalopod, or training it by rewards to perform a certain task?

The octopus is a prime candidate for humans to anthropomorphize because it is so smart, so responsive, and yet so different as to be essentially unreadable. Critics of the Netflix documentary *My Octopus Teacher* pointed out that watching the film, it was hard to know whether it was always the same octopus interacting with the human, since it is virtually impossible to tell octopuses of the same species apart. Was it always the same octopus that starred in the documentary or a series of different octopuses? How many of the relationships with various octopuses recounted in Sy Montgomery's bestselling book *The Soul of an Octopus* were only the author projecting onto an animal? It's impossible for us to know.

Another argument stressed by Nueva Pescanova in defense of farming octopus was that it would serve as a means of reducing the pressure on the declining numbers of wild stocks, a decline that most observers agree will continue to worsen as demand rises. "The octopus is an integral part of Galician identity, and we believe aquaculture can help ensure its continued well-being," David Chavarrías told me at O Grove. "If you know how to do it, always taking care for the well-being and sustainability of the species, why not do it? The animal may not be in imminent danger of extinction, but its numbers are declining. Aquaculture can serve to sustain the species while providing a high-quality food."

However, the belief that the aquaculture of a popular species serves to reduce the pressure on its wild stocks is by no means certain. A North Carolina State University study published in a 2019 issue of *Conservation Biology* looked at almost fifty years of data on wild fish and their aquacultured counterparts and concluded, "We found that aquaculture has expanded production, but does not appear to be advancing fishery conservation. In fact, aquaculture may contribute to greater demand for seafood as a result of the social processes that shape production and consumption."[29]

Opponents of octopus farming believe that producing more octopus, and making them available at a lower cost, could well increase demand for both the farmed and wild product and serve to accelerate scarcity in the oceans. The market for octopus in Europe and the United States currently serves mainly upscale consumers, and the cephalopod is generally priced out of the reach of most people. If the price comes down, thanks to more supply due to farmed animals, midrange markets could open up, greatly increasing the number of people potentially buying octopuses and augmenting the demand for them.

In addition to concerns about possibly increasing demand, the North Carolina State study recommended that if people were going to farm marine species, aquaculture should be used to meet the world's food needs, rather than as a way to make outsize profits:

> Socially prioritizing producing food (and in this case seafood protein) as a basic right to meet needs, rather than as simply another commodity in the global economy, and regulating production in an ecologically sound manner, would advance conservation goals while meeting human needs. This would require strong political-economic initiatives (policies) on national and global levels that better plan production, and implement, and enforce regulations that promote sustainability.[30]

Improving food security around the world, and increasing access to octopus by making it affordable as a possible high-protein meal for low-income households, does not appear in the marketing plans of any of the three most advanced groups currently trying to farm it. In Japan, the idea is simply to produce enough to meet current national demand and reduce octopus imports. In the first quarter of 2023, Japan imported more than ten thousand metric tons of octopus, according to the UN's Food and Agriculture Organization (FAO).[31] In Sisal, the plan is to produce a gourmet octopus that would keep a modest, steady flow of cash coming to Sisaleños from the high-end Mexican restaurants buying the cooperative's product to sell to well-to-do diners; and in Galicia, Nueva Pescanova wanted to make its mark in the global market, which was pretty much bound up with the idea of octopus as an upscale meal for those willing to pay an elevated price.

O Grove itself looked like it could use a little food security. The day I visited the marine research center was cold with a chilly drizzle, and after my interview with David Chavarrías I drove into the village proper. The tide was out, and the exposed tidal mud flats had dozens of people in worn rain slickers and rubber boots out digging for clams in the muddy suck, and a lot of them were not young. Later, at the fish and seafood auction shed, people brought in buckets, bags, and baskets of clams to sell, leaving with twenty-five or thirty euros for their efforts. It did not look like work anyone would do if they had anything else.

Behind the mud flats, and connected by a bridge to O Grove, was La Toja, a small, two-hundred-acre island with meticulously groomed estates belonging to some of Galicia's richest bankers and businesspeople. Until the mid-nineteenth century the island served the residents of O Grove as a place to cut timber, grow crops, or graze their animals. Then a spa was built to take advantage of the island's thermal mud and hot springs. Currently, La Toja and its luxurious homes cater to an upscale tourist trade with high-priced hotels, spas, a golf course, a marina, Spain's first casino, tennis courts, and a shopping mall. It was a pretty sure bet that none of the people digging in the cold, wet mud flats lived on La Toja.

Less than a mile away from the gated homes, the brood stock for Nueva Pescanova's farm-to-be was biding its time in the research center's shadowy basement. The company was jealously guarding those fifth-generation cephalopods against the day when they had a functioning farm and the octopuses could get to work reproducing. Nueva Pescanova had originally announced its intention to put its first aquacultured octopuses on the market in 2023, but in 2026 it was still waiting for the necessary environmental permits from Grand Canary Island, which the company continued to maintain would eventually be approved. If so, animal rights groups have said they are prepared to go to court to block the farm.

Opposition had intensified in March 2023 when an internal Nueva Pescanova document, outlining detailed plans for cultivation and slaughter, was leaked to the BBC. Among other things, it revealed that the manner of slaughter would be immersion in freezing "ice slurry" baths, a killing procedure occasionally used in the aquaculture of fish. This method, according to some experts, can be painful and slow.

Octopus fishers generally stab their catch once in a spot beneath the head where the arms join, and the animal dies instantly. Of course, if you have a thousand tanks full of octopuses and are factory farming them, handling and stabbing each one is far more time-consuming and inefficient in the long run than tossing them in frigid water and waiting for them to die. Peter Ulric Tse, an octopus neuroscience researcher at Dartmouth College who used octopus to study the neuroscience underpinning the concept of free will, told the BBC that using an ice slurry amounted to unnecessary cruelty. "To kill them with ice would be a slow death. It would be very cruel and should not be allowed."

With ground still not broken for the farm in 2026, Nueva Pescanova had yet to submit an acceptable environmental plan. The original Environmental Impact Assessment was rejected by the Canary Island government. "Nueva Pescanova's environmental report was inadequate, lacking basic information to allow the government to assess the impact of the farm on the environment and public health," environmental lawyer María Angeles López Lax told *Food & Beverage Insider* in 2024. "It's up to the company to prove that the farm would not impact on protected species or risk public health before permission can be granted, yet the company has failed to address even the most basic of these concerns."[32]

The company had things other than an octopus farm to worry about. In the spring of 2023, Nueva Pescanova itself had been put on the market. The fifth-largest seafood company in Europe, it had a checkered financial past and pressing financial issues in the present. Originally founded as Pescanova in 1960, it grew into an iconic Spanish brand of canned and frozen seafood, the largest seafood company in Spain, and the tenth-largest in the world. It was one of Galicia's economic mainstays, closely aligned to politicians from the conservative Partido Popular (Popular Party), which has governed Galicia for decades.

The Galician who founded Pescanova in 1961, José Fernández López, passed the company's reins to his son, Manuel Fernández de Sousa, in 1980. By 2010 the company had a presence in nineteen countries, with over ten thousand employees, and more than fifty fishing vessels at sea, bringing a constant flow of product to the company's processing plants in the Canary Islands. It had become one of Galicia's two top companies, behind only Inditex with its Zara fashion chain.

However, a financial earthquake struck Pescanova in 2012 when an audit revealed that there were two sets of books and that the company had over $3 billion in unreported debt. In addition, Fernández de Sousa had sold well over a million of his own shares in the months before the scandal broke. He resigned, the company went into the third largest bankruptcy proceeding in Spain's history, and in 2016 Fernández de Sousa was found guilty of fraud.

He exhausted his appeals in the spring of 2023 and began serving a six-year prison sentence in Madrid's Soto de Real prison.

There was no shortage of investors interested in rescuing Pescanova when it declared bankruptcy in 2013, recognizing its potential to return to profitability. CaixaBank, headquartered in Barcelona, expressed a willingness to assume the company and its debts. Instead, Abanca, a Galician branch of Venezuela's largest bank, Banesco, acquired a majority interest in the bankrupt company, saving it from falling into Catalan hands and earning the support and gratitude of Galicia's politicians.

The Galician struggle for self-sufficiency and liberty from Catalan exploitation of its marine resources stretched back to the eighteenth century and the critically important salt trade, which itself had been going on since the Romans colonized Galicia. The first Roman colonists arrived around 150 BC, and the region was incorporated into the empire around 30 BC. The Romans came for gold, but they stayed for salt, which was highly valued as a means of preserving meat, fish, and seafood. The ruins of Roman salt factories dot the Galician coast.

Two basic methods were used for conserving fish and seafood. One was to dry it, and the other was to salt it. Over the centuries, and long after the Roman Empire withdrew from Galicia, drying became the preferred method for conserving octopuses during the Middle Ages. Then, in the mid-eighteenth century, when word spread through the Catalonian fishing infrastructure in Barcelona, Girona, and along the Mediterranean's Costa Brava of the tremendous numbers of sardines off Galicia's Atlantic coast—as well as the quantities of salt that could be obtained there—thousands of Catalans came, drawn by a chance to exploit conditions in Galicia for their own benefit.[33]

Catalans have a reputation in Spain as perspicacious, hardworking, and aggressive businesspeople, adept at seizing economic opportunities. They brought to Galicia, from the Mediterranean on the other side of Iberia, advanced techniques for both fishing and conserving the catch with salt, which could then be exported to other regions of the country and European cities. The Catalans built boats to fish and plants to process and salt the sardines, and soon the eighteenth-century fishing industry on the Galician coast was virtually entirely controlled by *formentadores*, as the Catalans were called.

The locals who the Catalans employed supported their employers, but many in the Galician fishing community opposed the Catalan presence, claiming that Galicia's natural resources were being exploited to enrich Catalans. In 1774, José Andrés Cornide wrote, "The industrious Catalans . . . spread out through various groups of seafarers and trafficking along the coast ruin the fishing and the work of natives leaving them in a precarious subjugation, abusing the naivete of the crews with contracts that cause their ruin."[34]

It appeared for a while following Pescanova's ignominious bankruptcy that history would repeat itself, and once again astute Catalans would be helping themselves to Galicia's ocean-based wealth. When Abanca appeared and beat out the Catalan competition, taking control of Pescanova in 2015, the acquisition was helmed by Abanca's CEO Juan Carlos Escotet Rodríguez, who was ranked in 2023 by *Forbes* magazine as the fourth-richest person in Spain, worth over $4 billion. He held dual citizenship from Venezuela and Spain, lived in Galicia, and had the wholehearted support of Galician President Alberto Nuñez Feijoo of the autonomous region's ruling Partido Popular (PP). The bank quickly renamed the company Nueva Pescanova.

Abanca's investments have become increasingly important to Galicia's economy, and in 2024 the bank and its various affiliates were responsible for 16 percent of the region's gross domestic product, employing over 130,000 Galicians, according to the news agency Europa Press.[35] The bank's 2023 profits of over $780 million were more than triple those of 2022. By 2023, Abanca's share of Nueva Pescanova had increased to 97.5 percent, and the company had recaptured its position as Spain's largest seafood business, with sixty boats fishing in the Southern Hemisphere and annual sales of over a billion dollars.

In April 2023, Abanca announced that it was in exclusive negotiations to sell Nueva Pescanova to the Canadian seafood giant Cooke, which was prepared to pay a reported €150 million. In late July, however, Cooke drastically reduced that offer to about €50 million, based on due diligence by an external auditing firm, which concluded that Nueva Pescanova's debts and infrastructure deficiencies were greater than initially reported. Cooke's reduced offer was rejected by Abanca, negotiations were postponed until fall, and talks were not subsequently taken up again.

July of that year brought more bad news for the company with the release of Nueva Pescanova's revenues for the fiscal year that ended March 31, 2023. The company had lost €53 million, in sharp contrast to the €80 million profit of the preceding year. Nueva Pescanova's heaviest 2022–23 losses were down to its aquaculture operations, mainly shrimp farming, responsible for a deficit of €36 million. The downward trend continued with the company reporting losses of €73 million for the fiscal year ending in March 2024.

Juan Carlos Escotet announced in late October 2023 that the bank was no longer in a hurry to sell Nueva Pescanova and would be holding on to it "until the most opportune moment."[36]A sale to Cooke was no longer a possibility. In the annual consolidated report detailing the losses of 2023, the bank stated that it was still open to the sale of some part of Nueva Pescanova, as long as it remained in Galicia. Abanca has consistently declined to comment on the status of plans for an octopus farm, responding to my email queries: "We do not comment about our affiliated businesses."

Even in the event that Grand Canary Island granted the long-awaited environmental permits for the farm and a buyer for Nueva Pescanova was eventually found, it was not a given by any means that a new owner would be in a hurry to spend the estimated two and a half years and $75 million it would take to construct the farm, and to do so in as uncertain and unproven an aquaculture as octopus. And if a buyer is not found for the seafood giant, and Abanca holds on to it, the bank may not be willing to spend that much money gambling on octopus farming.

In early 2024, a Nueva Pescanova spokesperson told me on the phone that the company was still only waiting for all permits from Grand Canary to be approved before beginning farming operations. However, Silvia Fernández, a journalist in Las Palmas, wrote in the fall of 2023 that although Nueva Pescanova insisted publicly that it was going ahead with plans for the farm and only waiting for the environmental permits, unnamed sources in the company had told her that the company's financial setbacks, plus the fact that too many octopuses were dying at the center in O Grove, meant that plans for a farm were being abandoned.[37]

The director of the center, David Chavarrías, denied that octopuses were dying at an elevated rate in the Biomarine Center's tanks. "I can tell you that here we are continuing our development of cultivation in a totally normal manner, and I don't know from where came this rumor of massive mortality," he wrote me in an email.

By 2025, word had it that the company had notified the Grand Canary authorities that they were rethinking the project and that the application for the environmental permits was no longer pertinent. No one from the company was willing to confirm that, but it looked increasingly probable that Nueva Pescanova's octopus farm was on the verge of joining the others that had been announced as being just around the corner—and then failed to materialize. In late February 2025, David Chavarrías answered my email query as to whether the project was still in progress: "We are continuing work, and what we have to do now is define our next steps."

. . . .

Thousands of years ago, Galicia was home to as many as five thousand Celtic villages so isolated that each virtually constituted its own tribe.[38] This is still a characteristic of Galicia and its inhabitants. The entire region is dotted with small villages and a few large cities, its population dispersed among many small towns. Galicia is "a kind of enclave, an Atlantic bubble encrusted in a Mediterranean environment," according to the Galician writer Miguel Anxo Murado, who describes the dispersion of the population into its small villages as a singular characteristic of Galician identity in his book *Otro Idea de Galicia*.[39]

Records from 1752 show that in the Galician inland parish of San Xoán de Arcos, out of ninety heads of families whose jobs were listed, among the day laborers, tanners, and cobblers, nine were so-called *pulpeiros* who prepared and sold octopuses, presumably bought from the mule carts coming from the coast.[40] The ancient profession of arrieros who bumped their ways along roads dry and dusty in summer, or rutted and muddy in winter, driving their mule carts inland from the coast with dried octopuses to sell, lasted until the 1930s before finally disappearing completely. But, of course, what disappeared were the mules, wagons, and muleteers, not the octopuses. Motorized transport took over from the mule-drawn wagons, and that evolved into freezer trucks until, beginning in the 1960s, octopuses were caught at sea by ships that froze their catch onboard and brought them to port to sell in bulk to octopus wholesalers who might have thousands of frozen kilos on hand at a given moment. In short, Galicians continue to buy and sell octopuses as they have for centuries.

Almost all the octopus eaten by Spanish diners is prepared in the same fashion: *polbo á feira* (*polbo* being "octopus" in Galician), called *pulpo a la gallega* in the rest of Spain. The octopus is boiled whole and served in thin slices on a wooden platter, the slices dressed with a drizzle of virgin olive oil and a sprinkling of paprika, accompanied by a chunk of crusty bread. That's it, simple as can be. A bed of sliced potatoes is optional. In Galicia, a delicious variant has introduced melted cheese on top of the octopus. The cheese is usually *la tetilla*, a Galician cheese so called because it comes in the shape of a mounded, rounded, female breast. Traditional, mild, and creamy, la tetilla appears in written descriptions of Galician cheeses as early as 1753, and it makes an excellent topping for octopus.

Galicia has a centuries-old tradition of fairs and festivals, called *feiras*, often held in conjunction with the local marketplace. In 1999, the Galician anthropologist José Antonio Fidalgo Santamariña wrote about the food festivals: "Just in the province of Ourense we can count some 40 gastronomic celebrations, distributed over the length of the year, more than any other kind of celebration of a religious, secular, political, cultural, sporting or commercial nature."[41]

Polbo á feira derives its name from how octopuses are served at these fairs. Virtually every week of the year some Galician town or city hosts a feira. For people selling octopus to the Galician public, the modern-day pulpeiros, these fairs constitute an important source of revenue, because the food traditionally on offer is octopus. In the larger celebrations, over the course of a weekend, thousands of kilos of octopus are consumed. Pulpeiros go from fair to fair, carrying on one of Galicia's oldest and oddest occupations.

The pulpeiros and pulpeiras who continue to work these festivals become fewer and fewer as the years go by, younger generations finding easier ways to feed their families than to travel to strange towns and spend weekends boiling and slicing octopus. But pulpeiros haven't disappeared yet, and the movement of octopuses inland from the coast is best exemplified by the most important of these festivals, the Festa do Polbo, held every August nearly an hour and a half inland from the coast, in the town of O Carballiño. Each year, some seventy thousand people attend the octopus festival during the second weekend in August, most of them people coming from other places to eat their fill of polbo á feira.

Some pulpeiras choose to practice their skills in one place rather than follow the feiras. One such person was sixty-year-old María Luisa Rodríguez López, who had been preparing and serving octopus since 2000 at Casa Fidel O Pulpeiro, the restaurant she owned with her husband in Pontevedra. The pulpería was begun by her in-laws in 1956. She and her husband took it over when the in-laws retired, and it was her mother-in-law who taught her the traditional way of preparing octopus. That's who Luisa credited for making the restaurant a success—such a success, in fact, that it expanded to a second location across the street to accommodate the tourists and locals who crowded in at mealtimes. In the busy season of summer she boiled, sliced, and served over two hundred pounds of octopus every day.

The first thing that a diner saw on entering the restaurant was a gas ring beside the entryway, on which sat a big pot with boiling water and a whole octopus inside, its skin a scalded pink. Luisa stood beside it, keeping an eye on the octopus. She did not time how long the animal was cooking but relied on a visual assessment to know when it was ready to come out of the pot. "My mother-in-law was a real master; she had been cooking octopus since she was eleven years old. I've gotten better year after year," Luisa told me. "The way we serve it, sliced with just olive oil and a little paprika, it doesn't have sauces, or condiments, and there's no way to hide anything. It's delicate and exquisite, and people frequently tell me that they have never tasted octopus as good as what we serve."

It was, indeed, delicate and exquisite, and while I was in Pontevedra I didn't bother to eat octopus anywhere else, although there were dozens of places to do so, ranging from small cafés to opulent restaurants. Pulpeiras will tell you that although they may have to make do with octopus brought from North Africa, the local Galician octopus is the best there is. A local artisanal octopus fishery persists in Galicia, but the octopuses from Galician waters were too scarce and expensive for high-volume pulpeiras like Luisa. The octopuses she served came from North African waters. Regardless of their country of origin, they always came from the Atlantic.

While octopus landings in Galicia diminish every year, Galicians still fish for them, as they have for thousands of years, in the *rias*, the tongues of saltwater extending into the mouths of rivers along the coast. In coastal towns like Bueu, at the mouth of the Ría de Pontevedra, the remains of a salt factory from around 150 AD were found next to a Roman villa where the *salazon*'s owner presumably lived and close by another factory that made amphorae, the ceramic vessels that Romans used for storage and shipping. Fish and seafood were conserved with salt and could be shipped in the amphorae.

Today, the preferred way for artisanal fishers to capture octopuses is with traps, which are thought to have been used for capturing fish in the area since the eleventh century, but it was only in the early 1970s that octopus fishers began to use them. Before that, they fished with lines and hooks. These days, someone walking along the docks of any Galician coastal port is likely to pass piles of the octopus traps, *nasas de polbo*, waiting to be loaded onto boats. In 2019, some fifteen hundred artisanal fishing boats were licensed in Galicia to fish for octopus with traps. Each boat was allowed to set three hundred. The boats had to be under forty feet long, and the traps were pulled every day, according to regulations.[42] That is the legal part of Galicia's octopus market, but it is generally agreed that another market exists, supplied by both an illegal commercial catch and recreational fishers who sell to restaurants.[43]

Some eighty licensed boats go out in season to catch octopus every day from Bueu, according to José Rosas, the president of the town's *cofradía*, the fisher's guild. When members bring their day's catch back to port, they sell the octopus to the cofradía, which in turn sells it on along the market chain. José Rosas had fifty years of experience fishing, having worked on board big boats for years, going after cod in waters off Norway, Greenland, and Boston, then coming back to Bueu and fishing for octopus from his own boat. He knows full well that it is not easy to support a family by fishing, and it never has been. The problems are perennial: weather, fluctuating prices, trawling factory ships reducing the catch, and government bureaucracy. José Rosas had to worry about all those things and more, but one thing that didn't bother him was potential competition from octopus aquaculture. He was not concerned that Nueva Pescanova's farmed octopuses might soon appear in big numbers on the market, but he conceded that they could eventually be strong competition.

"They may have closed the [reproductive] cycle, but they haven't done so on a large scale," he told me in 2023. "Up to now, they haven't reached a scale where they could have profits that would make it worthwhile to advance with the project. They are there, coming along step by step, and they could eventually be a big competitor for us. Imagine that a client asks me for two thousand octopuses, each weighing three kilos. I can't do that. What we sell depends on the size we have on hand each day. However, aquaculture can do that, but even so we're not too afraid of them.

"We're one of the most important octopus fisheries in Galicia. The ocean's products that grow up and live in the wild have a different flavor, a different texture. For instance, it's not the same thing to eat a chicken raised inside as one raised outside. Everyone likes to eat something that tastes good, and for that reason our product is guaranteed to sell. For that reason, it would be important for the government to assure that the product is well ticketed and its origin is evident. If that's done, we're not worried, even though farmed octopus would be serious competition."

One advantage that fish farmers have over artisanal fishers is that they can count on a fairly stable production, year in and year out, a luxury that those going out in boats do not have. In the case of wild octopuses, as with many kinds of bounty from the ocean, some years are good and others are bad. This holds true for octopuses in both the Gulf of Mexico and the Atlantic Ocean, and it depends on many factors, some of which are known while others remain a mystery. "The octopus has a short life cycle," Rafael Bañon told me when we spoke at his home in Vigo. "It [the octopus population] has its ups and downs, depending on things like recruitment and oceanographic conditions, but the population is more or less stable."

He was a sixty-two-year-old marine biologist, former professor at the University of Santiago de Compostela, and was working with the Galician government as a consultant for marine science policy. He described taxonomy as his "passion." He was also someone who understood the kinds of impact that marine biology can have in the real world. When he completed his studies, he worked on fishing boats for ten years before finding work at the state oceanographic institute. He estimated that in Galicia, every day, six hundred of the licensed boats will be out fishing for octopus.

"They go out from dawn to sunset. They haul the traps in the morning. The boat will leave, for example, at six in the morning. They take the octopuses, put more bait in the traps and put them out. They [the traps] will stay there until the next morning. The boats don't go out on weekends, but the traps stay in the water."

Rafael Bañon has spent his life on the coast of Galicia, and he has been eating octopus regularly all his life. "It was much cheaper when I was a kid," he told me. "It was more accessible. Housewives bought it frequently. It wasn't a luxury like it is today. Now there are also a lot more markets—Japan and the United States. In my childhood, everything was for the market here. [Octopus] has been commercialized here since at least the fifteenth century. They dried it and sent it to Madrid."

Records from the end of the eighteenth century note that one of the ways that people fished for octopus was to tie a corncob on the end of a line and lower it to the bottom. Octopuses, intrigued by the color and shape, grabbed the cob and were pulled up. In the nineteenth and early twentieth century, the primary means of octopus fishing in Galicia was a weight on a line that ended in a pair of hooks. These were baited with crabs, and when an octopus grabbed a crab, the fisher pulled the line up with the octopus hanging on to the crab, refusing to relinquish it even as it was hauled out of the water. "The octopus is one of the most intelligent animals, but at the same time one of the stupidest," commented José Rosas. "It allows itself to be pulled right up beside the boat."

This two-hook octopus fishing rig was called a *rana* (frog), and it was not until the early 1970s that traps began to come into favor and replace them, according to Rafael Bañon. The traps also used crabs for bait, but currently most fishers prefer artificial bait, dry cakes made of fish parts, because they can be stored at room temperature, unlike fish and crabs, which must be kept cold.

With its abundance of fish and seafood, and its fertile land, Galicia might seem to have everything it needs to thrive. But the weather is cold and wet, and the land is rugged. When I visited in March it was gray; there were many days of a continuous, fine precipitation. It was barely rain, more condensed mist than drizzle, so particular to Galicia that it has its own name: *orballo*. In the mornings, fog lay halfway down the high hills, forested in eucalyptus.

In the sixteenth and seventeenth centuries, when the Hapsburg Empire dominated Europe with its fleet of wooden warships, the forests of Galicia were worth a lot of money. The original timber of Galicia was excellent hardwood, much in demand for the building of warships right through the eighteenth and nineteenth centuries, and nearly all of it was cut and sold. Construction of each wooden warship required some two hundred oaks and a large number of chestnuts, and over the course of those years the Galician forests were virtually destroyed.[44]

Today's forests of eucalyptus trees are not native but were planted by the regime of the dictator Francisco Franco Bahamonde to replace the old-growth timber his government cut down. Franco himself was Galician, born and raised by a drunken, philandering military father and a staunchly conservative Catholic mother in the small coastal village of El Ferrol. By the time the Franco government came to power after the Spanish Civil War ended in 1939, Galicia's native forests had come back. But cellulose was an early industrial aspiration of the Fascists, and to obtain it and ensure the rapid availability of trees, they cut down the existing forests, took the cellulose, and replanted Galicia with eucalyptus, a fast-growing eighteenth-century import from Australia. The eucalyptuses grew rapidly but depleted the soil, and because it's extremely combustible it created a higher risk of wildfire. It is an almost forgivable ecological mistake, given the pleasure of walking through the dark woods of tall eucalyptus, ground covered in the strips of bark these trees shed, the air heavy with their enveloping odor. But in the long run, eucalyptus is an invasive species and will prevent new forests of endemic trees from thriving. "The destruction of the native forests constitutes one of the most notable characteristics in the history of the Galician landscape," writes Luis Guitián Rivera, a geography professor at the University of Santiago de Compostela.[45]

. . . .

Despite the inherent difficulties of farming octopuses, Nueva Pescanova was not the only business in Galicia aspiring to do so. Álvaro Roura, born and raised in Vigo, started trying to grow octopuses in 2017. A thirty-eight-year-old marine biologist with a moustache, goatee, and long brown hair pulled back in a ponytail, he liked to surf the big waves off the Galician coast. "Octopus is Galicia's national dish, and naturally I ate them growing up. And they were delicious," he told me when we met for a beer in Vigo.

Roura was part of an early team of researchers delving into the nutritional requirements of octopus, and when he decided to work on how to grow them, he began by studying the behavior and nutritional requirements of wild octopus paralarvae and the journey they made before returning to coastal waters as juveniles ready to settle down to a benthic life. He figured that once he understood this, he could construct a successful cultivation process.

"They [paralarvae] are tiny, tiny creatures. We had to develop a technique to extract their stomachs. When they are born they enter the currents, and in four days they are 750 kilometers out to sea. They form a part of the plankton. But, finally, it's a willed behavior, and that's the magic of it, because, for instance, in these same conditions the larvae of squid don't do that. They stay close to the coast.

"[Paralarvae] behavior is very surprising. No more than born, the octopus takes such a risky journey. There are a huge number of animals that come every night to feed on the plankton where the paralarvae are. And then the paralarvae come back to the coast as juveniles. And that's the part we still haven't got. Do they come back on the surface? Or are they down near the bottom?

"The largest [paralarva] we found in the plankton had twenty-six suckers [on an arm]. The smallest had three suckers. We went out looking for one like that, and I had the good luck to find it in the plankton where there are always millions of animals. It was an incredible moment. You find one octopus paralarva for, like, every ten thousand animals you look at. I was discovering things about octopus larvae that no one in the world knew, and I decided to take what I'd learned about wild octopus and apply it to aquaculture."

In 2017, Álvaro Roura found some Galician investors, formed a business called Octolarvae, and in two years he raised two generations of octopuses. Then the pandemic struck, and by the time it was over, he had lost both the octopuses and his investors. When we spoke in the spring of 2023, he told me he had found some new funding and was putting together the basic infrastructure for an octopus farm. "The objective of the company I'm in now is to have an octopus farm. This is our mission for the future, but we're not in a hurry. It's a real challenge, and that's what I like. We're just three people. My wife is one, and she has an extraordinary sense of the octopus. She knows by looking when a tank is going well and when something's wrong. The third person is my colleague, Miguel, and we're doing the magic of growing octopus."

These three people were trying to do essentially the same thing that Spain's largest seafood company was spending huge amounts of money to accomplish, but that didn't worry Álvaro. "I don't think of Nueva Pescanova as competition. They are a multimillion-euro company and I'm starting from zero. They are gigantic and the researchers there have been working for decades. We're just beginning."

Álvaro Roura considered himself an environmentalist, but he had no problem with the idea of raising octopus for human consumption. He felt that those who objected to projects for farming them were not thinking the question through. "I think they're using a double standard. Pigs, chickens, trout, and salmon are all intelligent, and we eat them. Human beings eat animal protein. I understand people's concerns, and you have to treat them [octopus] in the best possible manner. If you don't raise them with caring, they'll die. They have to be farmed with a high quality of life."

To enjoy a good quality of life, an octopus has needs particular to its species, including a den of its own and plenty of sensory stimulation. It also requires a steady supply of nutritional food, and this is a problematic aspect that octopus farming shares with the aquaculture of other species. The question of feed is one of the most important decisions facing any aquacultural enterprise, a decision that will affect the fish farm's economic results, the growth rate of its animals, its immediate environment, and even global marine ecology. One persistent problem with farming various species is that farmed fish and seafood thrive best when they get their protein from a diet of other fish or crustaceans.

As aquaculture has become the fastest-growing sector of the food industry, the demand for an effective feed that makes ecological sense and also promotes growth has become increasingly important. In 2020, at the same time as the numbers of fish in the wild were declining under the pressure of heavy fishing, 20 percent of the catch from the annual landings of this increasingly scarce resource went not to feed people but to be turned into fish oil and fish meal with which to feed farmed fish and seafood. The major consumers of this feed were the farms that raised salmon, trout, and shrimp.

As it became evident that this was unsustainable, the industry's attention turned to krill. This tiny planktonic animal, which appears in huge numbers in oceans across the world, has a critical role in marine ecology: It makes up a large part of the diets of marine animals higher up the food chain, including whales, seals, other fish, penguins, and squid. The equilibrium of the world's fish population depends on the availability of krill. Almost half a million metric tons of krill were fished by humans in 2020. That is an immense amount of krill, and almost all of it went to aquaculture.[46] This, too, is obviously unsustainable.

For anyone who is able to eventually farm octopus, whether on the scale of Octolarvae, Álvaro Roura's tiny company, or Nueva Pescanova's million pounds a year, one of the primary considerations will be what to feed them so they will grow and thrive. Octopuses do not like most kinds of fish, and they generally want their food to be alive. Ask an octopus what it would prefer for supper and it would probably answer: a live crab. That's all well and good, but the problem is that crabs become expensive. Many of the cheaper things that an octopus will eat, like artemia, do not provide enough nutritional requirements for rapid weight gain. And in the battle to gain eventual public approval, octopus farmers will have to be able to show that they are feeding their charges in an ecologically sustainable manner.

One promising solution to the question of a sustainable aquaculture feed is insects. An insect-based feed has a number of advantages, according to an article in *Aquaculture Economics and Management*: Insects provide substantial nutritional value; their use has less environmental impact than others types of feed; and insects can help form a circular economy because they can be fed on the waste from other parts of the farming sector and are easy to raise. In fact, EU regulations allow for the use of some insects in animal feed, while in the US only the black soldier fly is approved. Insects have been shown to be a healthy source of protein in animal feed, but the principal problem discouraging their use is consumer acceptance.[47]

Research on the acceptance of new foods indicates that overcoming the barrier of habit is a primary concern. Studies on using insects as aquaculture feed are few, but they indicate that the older a person is, the less likely they are to eat fish fed on bugs, whereas younger people attach a higher value to the environmental sustainability of insects as feed and are more willing to try a new product.

Whether insects can be converted into octopus feed is something Carlos Rosas is exploring. "I don't know if it will work. Insects as feed provide low levels of polyunsaturated fatty acids. Octopuses have a low percentage of fat—less than 20 percent—and that's why they are such a healthy food. Maybe fifteen percent of that is polyunsaturated fatty acids, but perhaps using insect meal you can compensate by adding polyunsaturated fatty acids from things such as the waste parts of fish, like salmon and tuna."

Despite the difference in the scale of their endeavors, or perhaps because of it, a considerable amount of information about feed has been exchanged between Sisal and O Grove. The Moluscos del Mayab cooperative has even received a small amount of financial support from Nueva Pescanova, a grant that went toward providing a solar energy installation for the new farm. Carlos Rosas was flown to O Grove to consult with Nueva Pescanova about the feed he had developed in Sisal, and researchers from O Grove have visited the Octopus Project.

Pedro Domingues, a Portuguese marine biologist, worked on the team at Nueva Pescanova that closed the octopus's reproductive cycle. Domingues came to Nueva Pescanova while investigating cuttlefish biology, and he moved on to octopus research. He went to Sisal to do some work on shrimp aquaculture, met Carlos Rosas, and was fascinated by the Octopus Project. They began working to develop an economically feasible feed that would satisfy an octopus's nutritional requirements, and Domingues eventually took a job at the Sisal installation to continue his research. "The feed that we've developed here is giving excellent results," he told me when I spoke with him in his Sisal office in 2023.

Carlos Rosas and Pedro Domingues hoped to come up with a feed made from fish waste generated by the existing wild catch, things like heads, tails, guts, and skin. They had made progress, Pedro Domingues told me, but they were still trying to perfect their formula. It would be tremendously cost effective for a company like Nueva Pescanova, which had a fleet of fishing ships and processing plants, all of which generated a large amount of fish waste. Regulations prohibited ships from disposing of this waste at sea, so they had to freeze it and bring it back to land to get rid of it.

If fish waste could be used to make feed for octopus aquaculture, it would also benefit artisanal fishers who could earn extra money by selling waste they would normally throw away. "We are going to make the fishing industry a more efficient industry at the same time that we improve octopus production," Carlos Rosas told a *New York Post* reporter in March 2022.[48]

By the time I returned to Sisal in late 2023, fish waste was being combined with vitamins and minerals and shaped into pellets, replacing the paste we had spent such a large part of each workday putting into shells. "There is no other animal so capable of transforming poor-quality protein into high-quality protein as an octopus," Domingues told me. "The problem is that if you're farming them on a commercial basis, you can't be sure of a steady supply [of fish waste], depending on the catch. One day you may have a lot and the next day too little. Even if you have enough, you have to build warehouses to store it and have freezer capacity. You can produce a lot of octopus using fish waste, but it depends on the size of your commercial venture. In the long run, we still need to develop a feed."

In theory, by using fish waste as food, it would be possible to produce a tremendous number of octopuses, enough to help address the world's food insecurity with a protein source that reproduces in huge numbers, matures rapidly, and tastes good. What was once a poor person's food, and is now expensive, could help feed undernourished people around the world. Wars, extensive poverty, global warming, profiteering, and high prices for fertilizer all contribute to worldwide food insecurity, even though there is currently enough food produced in the world to provide everyone with the nutrition they need, were it well distributed.

If he is able to help develop a sustainable feed from fish waste, making octopus aquaculture viable, Pedro Domingues believed it could play a role in addressing food insecurity and provide enough revenue to pull people out of poverty in many places. "There's nothing more sustainable or easy than octopus," he told me. "They're pure protein, no fat, and if you have a fleet of fishing boats like Nueva Pescanova [producing fish waste], you would have the means for a very sustainable production."

A 2023 study published in the journal *Nature Food* looked at the nutritional importance of artisanal octopus fisheries to coastal communities in underdeveloped countries, focusing on the small-scale octopus fishery of Madagascar. One of the study's conclusions was that octopuses can provide necessary micronutrients in places where dietary staples may deliver energy to consumers but often fail to amount to good nutrition. Add nutrient-rich octopus to those diets, however, and people consume a number of key nutrients otherwise missing from their meals. David Willer, a zoology professor at the University of Cambridge and the study's lead author, comments, "Worldwide, nearly half of people's calories come from just three crops—rice, wheat, and maize—which are high in energy, but relatively low in key nutrients. Just a small serving of something very, very micronutrient rich, like octopus, can fill critical nutritional gaps. And, of course, if you get better nutrition as a child you're much more physically and mentally prepared for later life, which can lead to better jobs, better employment and better social development."[49]

Animal rights advocates read this and shake their heads. They are likely to agree that we certainly need to provide everyone with decent caloric and protein intake for their daily nutrition, but they insist there are other and better ways to do it than farming and slaughtering millions of octopuses. Even while silence reigns regarding any progress at Nueva Pescanova's factory farm, with not a single aquacultured octopus grown and sold since the original 2022 announcement, no one believes it is Galicia's last octopus-farming effort. Since 1995, somewhere in Galicia, someone, somehow is trying to farm octopus, and that is unlikely to change.

In March 2025 an email popped up in my inbox from the influential animal rights advocates Eurogroup for Animals about a seafood company called Profand, headquartered in Vigo. The company had nearly a billion dollars in annual sales of fish and seafood and had just received permission from Galician authorities to construct a state-of-the-art, 350-square-meter octopus aquaculture research facility in the Galician coastal town of Moaña. The company had some two dozen ships fishing around the world and aquaculture projects in Greece, Spain, and Ecuador growing fish and shrimp. Profand stressed that the octopus project's purpose was not to develop a commercial aquaculture but rather to do pure science, and they committed to a five-year investment in a research group with the goal of improving the viability of octopus paralarvae.

The email from Eurogroup for Animals told me: "Regardless of the company's intentions, such research could advance knowledge on the captive breeding of octopuses for future aquaculture purposes—an unethical and unsustainable practice that would have devastating impacts on these unique, intelligent creatures. . . . Octopuses are solitary, playful and inquisitive creatures, and high welfare octopus farming is impossible. The EU should not allow corporations to continue researching how to breed these animals in captivity."[50]

The research group named to do the investigative work for Profand's five-year project was Álvaro Roura's Octolarvae.

In Galicia, the search goes on.

Chapter 3

JAPAN

THERE is a Spanish expression used to describe a person who is totally befuddled and confused: *Más perdido que un pulpo en un garaje.* "More lost than an octopus in a garage."

It described exactly my state of mind when I got off the train from Haneda airport at Tokyo's Shimbashi subway station. As I wheeled my suitcase through the vast station, I began to think the trip might have been a grave mistake, even though I knew perfectly well why I had come to Japan: Per capita octopus consumption here is the highest in the world.[1] The animal is so popular that it is eaten as street food and is widely available, consumed regularly across the country as it has been for millennia. Until recently, it was a cheap, nutritious dish to put on the table, and even now, though it has become relatively expensive, many Japanese eat it frequently and count it as a comfort food, familiar from the meals of their childhoods. The Japanese love for octopus has also meant that nowhere else in the world have people been trying to farm them for so long.

I knew that even after more than sixty years of research, time and money were still being spent on efforts to develop octopus aquaculture, although not so much as a single farmed octopus had ever reached the market. A steady supply of octopus was important to the nation. To do without it in Japan would be like not having beef in the United States. It stood to reason that if there was anywhere in the world where people would eventually farm octopuses, it would be Japan, and I had decided to go see for myself. The primary species captured there is *Octopus sinensis*, called *madako* in Japanese, or common octopus. It is so similar in shape, size, and taste to the Western world's common octopus, *O. vulgaris*, that only in 2016 was it recognized as a separate species, thanks to research by a British marine biologist and taxonomist, Ian Gleadall, who lived in Sendai, Japan, and was a university professor there for nearly forty years before retiring.[2]

I wrote to Carlos Rosas in Sisal and asked if he would like to go to Japan with me. He wrote back that he had always wanted to visit the country and experience firsthand the Japanese devotion to octopus, but, unfortunately, he could not get away. He recommended that I contact Ian Gleadall. When I did, the Englishman responded that for a minimal fee he would be glad to accompany me and serve as a translator. He wrote that he had commitments the first couple of days I was scheduled to be in Tokyo, but after that he would be free to travel around the country helping me learn about octopus in Japan.

As I stood in Shimbashi station like a small, still rock with a wide river of people rushing around me, it felt like I might not even last for a couple of days. A tight budget had motivated me to reserve those first two Tokyo nights in the cheapest lodgings I could find online—apart from dormitory settings—in something called a "capsule hotel," advertised as being a short walk to Shimbashi. I have rarely been as lost as I was the minute I got off that train from Haneda airport.

I was definitely feeling the effects of the fourteen-hour flight from Barcelona and was more than ready to lie down, but to do that I first had to find the hotel. Even before that, I had to figure out how to leave Shimbashi. I had drastically underestimated the size and complexity of a major Tokyo subway station. It was teeming with people, all moving quickly and with purpose. The station itself seemed to be the size of a small city, long corridors stretching in every direction lined with shops and multiple exit signs, none of which I could read. I wheeled my luggage aimlessly through the heaving crowds of people who knew exactly where they were going and gratefully ducked into an office labeled with the only English sign I had seen in the entire station: TOURIST INFORMATION. When I asked the woman behind the counter if she spoke English, she answered in the affirmative, but her English was a far cry from mine. She looked a little surprised that I was searching for a capsule hotel but nevertheless gave me to understand that my destination was four blocks from Shimbashi, opposite a McDonald's. "You know McDonald's, yes?"

As crowded as the station was, the streets were even more so. I wheeled my suitcase up and down numerous blocks, in a state of fatigue and near panic, without seeing either the hotel or a McDonald's. The GPS on my cell phone seemed to have even less idea where I should go than I did. I passed lots of other hotels and stopped in a number of them to ask for directions, reasoning that the desk clerks would speak English. They did, after a fashion, but it was not a fashion that enabled me to determine where I was or where I needed to go. It was a warm day, and by the time I found the capsule hotel half an hour later, and only one block, after all, from Shimbashi station, I was sweating and exhausted.

The space an individual pays for in a capsule hotel is as close to a preview of what it would be like in a coffin as most of us are likely to have this side of the grave. The capsules are six feet long by three feet wide, with a curtain at one end to close for privacy. A short person like myself could sit up in them, but they were too low to stand. Bathrooms were communal, and suitcases went in a locker downstairs because there was no room for them in the capsule. Trains arriving and leaving Shimbashi sounded like they were rolling right through my tiny space, and earplugs were provided free of charge. Thankfully, there were drinks—soft or alcoholic—along with free rice and miso soup available all day in the lounge, a toothsome curry for breakfast, a pair of pajamas and slippers, excellent massage chairs, as well as access to an *onsen*—a large hot springs pool—in the basement. Despite the amenities, two nights in the capsule were about one night too many, and I was glad when the day arrived to rendezvous with Ian Gleadall and begin our journey.

With seven decades of research efforts behind them, I was certain the Japanese must be closing in on a successful octopus farm. The first serious attempt was made in 1962 when a group of researchers working under marine biologist Kouzo Itami at the Hyogo Prefectural Fisheries Experimental Station on the shores of the Seto Inland Sea obtained two hundred common octopus paralarvae from eggs. Their work was initially viewed as a promising first step toward a large-scale aquaculture, but their paralarvae had an unacceptably low survival rate. Ninety percent of them died before settling to the bottom and beginning their benthic lives, and the few that survived to settle as juveniles did not thrive. In the following decades research continued, but keeping at least a majority of paralarvae alive to a saleable size continued to prove elusive.[3]

Researchers did not give up. In 2001, for instance, an organization called the Japanese Sea Farming Association announced in a newsletter that it was set to "recommence" an experimental octopus farming program, having previously achieved a survival rate of 75 percent of juvenile octopus and reared them for twenty-five days. "Globally," the announcement concluded, "this was the first time this had been achieved. . . . However, most of them perished within three months."[4]

A quarter century later, Ian Gleadall told me when we rendezvoused at Tokyo Station to begin our trip, efforts were ongoing, although shrouded in secrecy. He was part of one such effort. "We're working on a whole bunch of things at the moment, which unfortunately I'm not allowed to talk about. A lot of research is going into it right now, but we're still not there yet. Not by any means. It'll be another few years before we see any commercial fruits."

At seventy-one-years old, Ian Gleadall was not remarkably tall for an Englishman, but he often seemed so when among Japanese. A solidly built Brit whose accent had not disappeared even after forty years in Japan, he had thinning, sandy-colored hair and greenish-blue eyes. He said he was still fit due to having practiced Thai boxing when he was younger. He studied marine biology in the middle of England in the mid-1970s at the landlocked Sheffield University, and one of his professors was John Messenger, who did groundbreaking research on cephalopod behavior.

Messenger's lectures awakened an interest in octopuses, and Gleadall began to specialize. For his hands-on experience, he spent summers working in Naples at the Stazione Anton Dohrn. As a student of John Messenger's, he was able to spend those summers working with some of the outstanding names in the field, people like John Z. Young and Brian B. Boycott. "I learned from the best," he told me. "Naples completely changed my life. It wasn't just the research. I was raised in a working-class, meat-and-potatoes, English family. During my time in Naples I discovered so much—different food and a different attitude toward living. God, how we ate!"

Not long after completing his doctorate, he was offered a job by the Japanese government to teach at Tohoku University in Sendai. "When they told me I'd be going to Sendai, I looked on a map and saw it was close to a place called Zao, which had a triangular mark indicating skiing, and I took that as a good sign."

After a few months in Sendai and still struggling with the Japanese language, he met his future wife, Etsuko, at a party. "She was one of the few people at the party who spoke English. When she told me she was from Zao, I asked her if she skied, and she told me she'd skied since she was a child. We made a date to go skiing."

They eventually married, much to the initial consternation of her parents. Etsuko came from a traditional Japanese family, living in the countryside in Zao an hour away from Sendai. They were not happy to learn that their daughter was going to marry an Englishman. "When we got engaged, it was traumatic for my in-laws," Ian told me. "Their immediate reaction was, 'Hold on, that's just not going to happen.' Then my wife talked to her mother, and she finally came around, and then that was it. Once that happened, that was it. As in a lot of Japan, her father thinks he runs the household; it's a very male-dominated society, but underneath it's the women who run everything. And, of course, when we had kids that changed a lot of their attitudes. I hadn't really thought about having kids. My father was kind of a cold man, and fatherhood didn't really appeal to me. It was not on my mind. But my wife insisted, and I'm glad. My two sons are the best things that ever happened to me."

As our trip got underway in the comfortable *shinkansen*, the bullet train, Ian described the current status of octopus. "Octopus has been overfished for a long time in Japan, and don't expect to find a specific answer as to why catch numbers are declining," he forewarned me. "It's just hanging on by a thread. The catch varies drastically from year to year. A good year depends on things like when the water warms up, or when crab larvae are available to eat, but lately it has been all bad years, and no one's sure why. You can ask the people we talk to on our trip, but you'll get a number of different answers and no one knows for sure."

Tako is the Japanese word for octopus, he explained, and the animal is deeply woven into the collective Japanese consciousness and cuisine. "A hell of a lot of octopus is eaten in Japan. All the octopus they can find gets sold, and because of that the price is high. It's going up and up. That's why they [the Japanese government] are so keen on getting aquaculture going."

Frozen and fresh octopus is always for sale in supermarkets, and pre-cooked octopus is usually on offer at Japan's many convenience stores for a moderate price. Take it home, cut it up, heat and serve. Or it is consumed in sushi, sashimi, with rice, in a salad, or in a variety of other dishes. Kids scarf down fish sausages shaped like a hot dog and called "octopus wieners" because they have eight cuts at one end allowing them to look kind of like an octopus—if an octopus were shaped like a hot dog. Octopus wieners are staples in the bento boxes that kids bring to school, which are the equivalent of North American lunchboxes. In Akashi, on the Seto Inland Sea, a local favorite food is, in fact, an octopus bento. Large pieces of octopus stewed in a soy-based sauce are put in a small pot on top of rice, along with vegetables in season, maybe some bamboo shoots, shitake mushrooms, and conger eel. Called *hipparidako meshi*, it is sufficient for an adult's lunch and is frequently eaten as such.

The tastiest octopus in the world, Japanese aficionados will tell you, are the madako, the common octopus, caught in the waters of the Seto Inland Sea some three hundred miles west of Tokyo, and that's where Ian and I were headed on the bullet train. This superior flavor of the sea's *O. sinensis* has traditionally been attributed to its strong tidal currents. The waters are relatively calm since it is an inland sea, essentially a vast bay some 280 miles long from east to west. But it has strong tidal currents, and the tide can rise and fall as much as twelve feet in a single cycle. This means, the theory goes, that octopuses must cling tightly to rocks on the bottom to avoid being carried away, which develops short, thick arms making for the most flavorful meat. In places like Osaka beside the Seto Inland Sea, people have been eating local octopuses for a very long time. Numerous old, discarded ceramic octopus pots appear in the nets of trawlers, and archaeologists have dated some of those pulled up from the bottom of the Seto Inland Sea as far back as the mid-Yayoi period, from 100 BC to 100 AD.[5]

The problem today is that there are not that many *O. sinensis* still swimming in Japanese waters, much less just in the Seto Inland Sea. Even adding in the other frequently eaten species, the giant Pacific octopus (*Enteroctopus dofleini*), the numbers are not enough to meet national demand. The giant Pacific octopus lives only in Japan's colder, northern waters, and like the common octopus it has been overfished. The total annual catch of both species is not nearly enough to satisfy Japan's octopus market, which annually consumes about seventy thousand metric tons.[6] Demand must be met by imports from other places.

The same trawlers that bring up ancient pots in their nets are often blamed for the drastic decline in the number of octopuses caught each year, both for the large numbers of animals they capture and the destruction of the seabed that their nets can cause. But landings are declining in an alarming fashion even in those waters closer to shore where trawlers do not work and where artisanal fishers still use pots and traps to catch their octopuses. These days they prefer plastic pots rather than the traditional clay ones because they withstand setting and hauling much longer than ceramic pots—and they are cheaper. So unanimous is this preference that there was only one manufacturer of ceramic octopus pots left in Japan. His name was Hisano Tomohiro, and he lived in Hofu City, a city of some 115,000 people on the far side of the Seto Inland Sea. Before examining the present and the future, Ian suggested we begin with the past—a visit to Hisano to learn how clay octopus pots used to be made, a vocation that was thousands of years old and disappearing even faster than the octopuses they were made to catch.

It turns out that the conversion to plastic pots has some worrisome environmental consequences because they do not degrade. Ceramic pots, on the other hand, can disintegrate in a few years, making them environmentally friendly though an expense for octopus fishers to replace. Today's pots contribute to the ever-growing load of plastic pollution in our oceans. But demand for affordable pots is high, and plastic answers the immediate purpose.

To visit Hisano, we took the shinkansen to Osaka where we changed to a local for Hofu City. As the train reached the Osaka station I began a long series of the kinds of embarrassing moments experienced by many a clumsy Western visitor to Japan. The young Japanese guy across the aisle from me on the shinkansen was well provisioned for his trip, eating rice balls wrapped in seaweed and drinking cans of Sapporo beer, all atop the folding table in front of him. When Ian and I got up to change trains in Osaka, I had to manage my gaijin tourist's burden: a backpack and a large suitcase. As the train pulled into the station and I stood up to wheel my baggage down the aisle, the corner of my backpack tipped his beer over, spilling it on his pants. I glanced over my shoulder in horror at the surprise and disgust on his face as he looked down at the pool of beer spreading across his lap and then up at me, while a rivulet of beer ran down the aisle and under the suitcases of those of us who were disembarking. I did not have the words to apologize, nor time to do so even if I had known how, because the shinkansen's stops were brief and the line of people waiting in the aisle to get off was moving right along.

Fortunately, Ian Gleadall was always willing to overlook my gaffes as a bumbling Westerner and to share his grand store of octopus knowledge with me. He had quickly satisfied himself that I knew how and when to perform the basic bow, which replaces a handshake. He explained to me how I had to have a small gift to present to every interviewee; cautioned me to always pass over my business card with both hands; and reminded me to show up for interviews wearing a tie.

Hisano Tomohiro had no trouble picking us out of the crowd when we got off the local train. He drove us to his factory on the outskirts of Hofu City. Across the road, I saw the Seto Inland Sea, shining blue and calm, the heart of the world's octopus fishery even if its stock was disappearing. In addition to creating a huge potential market for anyone who finally succeeds at farming octopuses, the steep decline in annual landings had provided Hisano Tomohiro with a windfall. While fishers have virtually stopped using clay pots, the government, alarmed at the shrinking octopus fishery, has undertaken a program to try to attract females into pots so they will have a safe place to lay their eggs, multiply, and be fruitful. This, reasons the government, is a good strategy to counteract the decline of the octopus population, and in late September, just before the female octopuses will begin to lay eggs, fishers are hired to toss ceramic pots overboard, hoping that they will provide a lying-in home for expectant mothers and their eggs. No one could tell me how many pots are actually being planted on the bottom of the Seto Inland Sea for this purpose, but they are enough that Tomohiro sells virtually all his annual pot production to the local government.

Four times a year, he fires up his four brick kilns, each of which is forty-two feet high and three feet across. To do so requires stoking the kilns' fires with wood until they heat up to about 2,500 degrees Fahrenheit. Each year he burns thirty-two tons of wood to get the job done. The process of firing the pots, which he has made from local clay, takes seventy-two hours, with smoke pouring out of a towering brick chimney around the clock. During the three days of firing, he has to continuously adjust the position of the pots on the floor of the kiln. Four times a year, the forty-two-year-old, round-faced, easygoing Tomohiro winds up exhausted, with fifteen hundred fired pots and all the hair on his arms burnt off. It takes a week for the pots to cool down. The kilns, built in a ramshackle warehouse on the outskirts of Hofu City, were originally constructed after the Second World War to manufacture clay drainpipes and were left to crumble when inexpensive aluminum and PVC pipes conquered the market.

"I was working as a wedding planner in Hiroshima and came to Hofu City with a group of friends one weekend a decade ago," Hisano Tomohiro told us. "I saw this place and was captivated by it. I moved here and worked at odd jobs until I learned how to do this and got this place up and running eight years ago. It changed my life. Now my parents live here, and I just got married this year."

One of the jobs Tomohiro did before beginning to turn out octopus pots was working in a neighborhood seaside restaurant, and he invited us there for lunch. The modest Hofu City eatery with eight tables and a local clientele was next door to a fish and seafood market, so the delicious lunch we ate could not have been fresher. We each devoured a dozen oysters grilled on a hibachi mounted on our table. After that first course we moved on to scallops, squid, giant snails, and mackerel. All of these Tomohiro cooked for us on the grill, and then we were served a bowl of locally caught octopus chopped up in rice. Later, we walked off the lunch, ambling along the port past piles of rope and plastic octopus pots. Sleek-winged kites wheeled overhead in the bright blue sky above the Seto Inland Sea.

Octopus in Japan does not mean an expensive plate of food prepared by a gourmet chef, as it does at an upscale Manhattan restaurant, nor a high-priced item at the local fish market, as in Barcelona. Even today, with prices rising, many Japanese routinely eat it as part of a meal, and it is no surprise when the answer to "What's for supper?" has octopus in the reply. Often that octopus will have been caught in the Seto Inland Sea. Our train back to Osaka from Hofu City traced the sea's coastline. Out the window toward the north, on the landward side, I could see high, green, forested hills surrounding the small, bright-blue bays we passed.

At one of the stops a short, middle-aged woman walked by us with a pack on her back, and a small bell on the pack rang softly as she stepped down the aisle. "Bear bell," Ian told me. "A lot of people come to this area to hike, and when you're hiking in the woods in Japan it's a good idea to let bears know you're coming. You don't want to surprise a bear."

He paused. "On the other hand, some people don't use one because they think a bell will tell a bear, 'Here's breakfast, come and get it.'"

Before reaching Osaka, we stopped for the night in the small port city of Mihara, where Ian had reserved rooms, thinking there would be a lot of octopus fishing activity. It was raining hard and already dark when we descended from the train to an empty parking lot. Fortunately, a taxi appeared, the female driver dressed impeccably in a black suit, like many Japanese taxi drivers, wearing white gloves and a surgical mask. The next morning, Ian rose at the crack of dawn to walk down to the port in search of octopus fishermen while I slept. He didn't find any, and it was on to Osaka, the second largest city in Japan.

Our first visit during our Osaka stay was to Akashi, a coastal city of some 326,000 people, less than an hour west by train from the big city. Akashi is the capital of Japan's extensive octopus world, famous across the country for its Seto Inland Sea octopuses. We had an appointment to speak with Hiroaki Ebisumoto, the union president of the Akashiura Fishery Cooperatives, which received and sold the catch from the eighty fishing boats belonging to its members at a daily auction. His office was upstairs, at one end of a long, covered shed by the sea.

The floor of the shed was kept covered in calf-deep seawater. Here, every morning, it was possible to buy over a hundred kinds of fresh fish and seafood, caught by cooperative members, everything from sepia to sea snail, eel to octopus, mackerel to turbot, all of them kept alive overnight in permeable plastic crates on the flooded floor so that they would calm down after the stress of being caught. The next morning they were sold at auction to a small crowd of restaurateurs and retailers who stood on metal risers above the wet, concrete floor. The fish, cephalopods, and crustaceans passed in front of the bidders, alive and as perfect as when they were caught the day before. As soon as they were bought, they were killed.

Hiroaki Ebisumoto was in a meeting when we got there on a sunny and breezy weekday morning, and we stood outside on the dock waiting to be summoned upstairs to his office. In front of us, an elderly, thin, white-haired man squatted on his heels at the dock's edge, fishing with a long pole, a bobber floating in the water a few feet in front of him. A tall and regal grey heron was perched motionless on a bollard a few feet from the fisherman. The elderly man was almost as motionless as the bird, as he sat crouched on his heels, back straight, an air of quiet around him, a posture I would not have been able to maintain for more than a few seconds. He appeared quite comfortable squatting there as trawlers slowly passed by in front of him on their way to the shed to drop off the morning's catch. He waited patiently, watching his bobber, perhaps hoping for a small fish, maybe something to take home for his wife to cook for lunch. Neither he nor the heron moved in the quarter of an hour Ian and I watched them, before Hiroaki Ebisumoto appeared at our side to bring us out of the wind to his office.

It was large and comfortably furnished with a wide desk, conference table, pictures of past cooperative presidents on the wall, and posters for Bruce Lee films. We sipped green tea. With a bow, and holding it in both hands, I presented him with the gift I gave everyone: a mosaic lizard refrigerator magnet in the trencadís mosaic style of Barcelona's most famous twentieth-century architect, Antoni Gaudí. Many Japanese know and love Gaudí, and Japanese funds had gone a long way toward completing his most famous work, Barcelona's monumental cathedral La Sagrada Familia.

Hiroaki Ebisumoto was a thin, small man, hair cut very short, in an open-necked shirt, wearing delicate-looking gold-rimmed glasses. He began fishing for octopus in 1981 and continued to do so until assuming the union's presidency in 2009. He had a self-satisfied smile when he talked about Akashi's octopus: "In this area, octopus is special to us. We catch a lot and we eat a lot. It has a great flavor, it's easy to prepare, and it's a very common food. You just take the insides out, tenderize it a bit, and boil it. I remember my grandmother cooking it, and she was quick. I mean quick. She didn't boil it for long. Four or five minutes. Longer than that and it begins to toughen up. It's easier than fish because it doesn't have all those bones."

The high regard in which octopuses from the Seto Inland Sea are held is no accident, he assured us. "The Akashi strait and the Harima-nada Sea that follows it are full of crabs, shrimp, and small fish that serve as foods. The octopuses eat a lot of high-quality food, and the fast-flowing waters around Akashi give the octopus plenty of exercise to hold on, keeping its flesh firm. Our octopus really does taste better. Akashi once had the largest catch of common octopuses caught in Japan, with a thousand tons a year."

Hiroaki Ebisumoto acknowledged that catches were way down and prices way up. Every year two or three cooperative members were giving it up due to the reduced catches. The last year Akashi had registered landings of a thousand tons was 2015, and since then numbers had fallen precipitously. Six years later, in 2021, the annual catch was only 143 tons. Numbers were slowly going back up, but they still constituted bad years. By 2023, 217 tons were recorded, and while 2021–23 were the lowest three years on record, Ebisumoto was convinced that the octopus population was on the rise and that the decline was simply a part of an ever-present cycle of good years and bad. "I expect they'll come back. You know, they only live for one year, so the numbers can quickly recover. We have had bad years before, then all of a sudden the catches will get a lot better."

The cooperative's boats fish with trawl nets. Ebisumoto maintained that far from damaging the seabed, it was like plowing a field, and the nets stirred up the bottom, actually freeing phosphates and nitrates that served as nutrition for plankton, which then fed fish, seafood, and octopus paralarvae. In fact, some researchers, including Ian Gleadall, thought part of the reduction of the octopus population could be a result of the cleanups of the rivers flowing into Japan's seas and oceans, which has reduced the amount of waste matter in them and subsequently the amount of plankton fed by that waste. Less plankton meant less nutrients for the paralarvae and fewer juveniles to grow into adults.

In the sixteenth century in Akashi, just as in Galicia nearly seven thousand miles away, octopus was so highly valued that it served as currency. For much of Japan's history, a feudal system was the prevalent economic structure, much as it was in Galicia, and in Akashi serfs were able to pay for land rental with dried octopus. "As ancient octopus pots are often found pulled up from the seabed, it is easy to imagine that this area was relatively convenient for people in antiquity to obtain octopuses to eat," Ebisumoto said. "The complex seabed topography and fast currents around the Akashi Strait, from the deep sea to the shallow, sandy areas, provide abundant nutrients, which are then used by the plankton to produce a wide variety of food organisms, including small fish, and this is why the Akashi area has become a natural fishing ground where octopus resources have been maintained and octopus fishing has always gone on."

Even with the cooperative's boats landing less, Ebisumoto was not worried about competition if octopus aquaculture became a successful commercial reality. "It's true that octopus has gotten more expensive, but it's not *that* expensive. People haven't stopped eating it. The price was a little too low over the past few years, and that's part of the reason it has gotten higher. People are going to keep eating octopus, and with aquacultured octopus it would be just like the case of other fish that are farmed today and sold in the market side by side with the same fish from the wild. I don't see aquaculture as a big problem for us. We'll be able to charge a premium for the wild octopus as they do with lots of other fish. Wild-caught salmon is more expensive than the aquacultured product, but lots of people are willing to pay the extra because it tastes better. Like salmon, people who eat octopus will always be able to distinguish between the tastes of farm raised and wild caught."

He laughed and nodded. "Japanese people will always be eating octopus. In sushi, sashimi, and particularly in *takoyaki*."

Takoyaki. The name refers to its principal ingredient, octopus (*tako*), and its preparation, cooked over direct heat (*yaki*). Around Osaka, virtually any conversation about octopus will include a reference to takoyaki, the city's iconic octopus dish. Takoyaki is made with wheat-flour batter into something the size and shape of a golf ball, with a small piece of boiled octopus tucked into its center, along with tempura scraps (*tenkasu*), pickled ginger, and green onion. A tasty, slightly sweet sauce and a dab of Japanese mayonnaise top each doughball. Special cast-iron griddles can cook dozens of takoyaki at a time, which are sold from storefronts or stalls facing the street, often with a brightly painted picture of an octopus painted on them.

A one-serving takeaway cardboard container holding six takoyaki, which includes an elongated pointed toothpick for spearing the balls, costs less than four dollars. It is cheap, tasty, and filling. Many private homes have electric takoyaki griddles in their kitchens, much as North Americans might have waffle irons. Prepackaged takoyaki and takoyaki batter are sold in supermarkets and convenience stores. While it has only existed since the mid-1930s—when, according to a widely accepted account, an Osaka street seller named Tomekichi Endo first made and offered takoyaki—it has become a national dish and accounts for a great deal of the octopus eaten every year in Japan.

One Saturday afternoon, Ian Gleadall's colleague and friend Hidetaka Furuya volunteered to take us on a takoyaki tour in Osaka. It started with a subway ride from the Namba station where we met him. Furuya was an Osaka native, but subway stations in Japan are so complex as to be barely navigable, even for natives—and he used Google Maps on his phone to move us to the correct subway platform.

Hidetaka Furuya was a biology professor at the University of Osaka's Graduate School of Science and famous among cephalopod researchers for his work on dicyemids, which are parasites of cephalopod kidneys. In describing his research and its applications he has written, "Because of its very simple body composed of 20 to 40 cells, dicyemids . . . serve as the simplest model system for the study of cell differentiation and morphogenesis in multicellular animals."[7]

In addition to spending his days examining those tiny parasites, he writes poetry. "I always enjoy writing," he told me. "I have loved poetry from Japan's ancient times since I was young, and after I retire from the university, I plan to write poetry while traveling. The poetry I write follows a style called 'waka,' a classical form that is a short poem and uses ancient words spoken over a thousand years ago. I dream of traveling across Japan, writing many poems inspired by the local landscapes and culture, and compiling them into a book."

Before we began our tour of the ubiquitous takoyaki stands along the city's streets, Hidetaka Furuya brought us to a large kitchenware store in the center of Osaka. A whole long wall was dedicated to kitchen knives: perfectly weighted, beautiful, long, gleaming blades, easy to imagine making the thinnest and most perfect of slices. Some of them were priced at hundreds of dollars. Close to the store's entrance, among the more plebian merchandise, a shelf displayed electric takoyaki griddles with rows of small copper-lined molds in which to pour the batter. They came in a variety of sizes, from griddles with only a half-dozen molds to larger ones that could produce twenty-four at a time, presumably for large families. A griddle with four rows of five molds, capable of twenty takoyaki, was priced at 3,900 yen (twenty-six dollars). Pour takoyaki batter into the molds, drop a small piece of boiled octopus into the middle of each while it cooks, keep turning the balls of batter until they are golden brown, add whatever topping you desire, and you've got home-cooked takoyaki. Unfortunately, even the sleek six-serving model was too heavy and bulky to bring back to Barcelona. Reluctantly, I left the takoyaki griddles on their shelf.

For now, takoyaki is not well known outside of Japan, but that is changing. For instance, it is available at an ever-increasing number of Japanese restaurants in Los Angeles, where the world's best baseball player, Shohei Ohtani, noted for both his superior pitching and batting skills, joined the Los Angeles Dodgers by signing a ten-year, $700 million contract at the start of the 2024 season. Ohtani, indisputably the greatest Japanese player ever to appear in the major leagues, was expected to bring tremendous star power to the team. One thing he also brought was takoyaki. For the first time, it was among the refreshments you could buy from the food stands at Dodger stadium, and it was the real thing. All its ingredients, from batter to octopus, are sent across the Pacific Ocean by Gindaco, a giant franchiser with some four hundred outlets across Japan featuring takoyaki.

Hidetaka Furuya took us to the Dotonbori district, named after the canal running through it and one of Osaka's principal tourist destinations. The main street was a neon wonderway, packed wall-to-wall with people enjoying this sunny Saturday afternoon, bright signs on both sides of the street advertising food stands of many kinds. Numerous large neon octopuses denoting takoyaki eateries were prominently mounted on buildings, and it seemed like every other little eatery in Dotonbori was a takoyaki stall. Many of them had long lines, people waiting and watching expectantly as the takoyaki maker behind the counter kept a constant motion, turning the balls with an awl pointed like an icepick. No sooner were they out of the griddle, topped, and in cardboard boats for the people at the front of the line than batter was poured into the molds for the next batch in a nonstop workday. The griddles at takoyaki stands need to be replaced every five years, Hidetaka told us.

We tried the takoyaki balls from three different stalls in Dotonbori. Each stall's takoyaki had a slight variation in the crispiness, or sauce on top, or amount of octopus at their centers. We were already full as ticks when Hidetaka Furuya told us he had saved the best for last. Generally speaking, there are two different styles of takoyaki: The Osaka variation is creamier and not so crisp, while the Tokyo takoyaki is crispier on the outside and less creamy inside. There is, however, a third kind of takoyaki, which is generally only found in Osaka, and then only at a limited number of places. It's called *akashiyaki* and is lighter than its cousins, shaped more like a biscuit, with a piece of octopus at its center and served without any sauce at all. Instead, it is accompanied by a small bowl of a thin fish broth called *dashi*, into which the akashiyaki is dipped. The standard wooden toothpick is not supplied, the idea being that the dipping should be done with chopsticks, so I was afforded yet another opportunity to embarrass myself and Ian Gleadall. I endured discreet, bemused glances from the Japanese around me as my chopsticks rolled between my fingers, and I dropped the whole akashiyaki into the dashi with a splash. It took me a few humiliating tries to get each piece dipped and into my mouth, but my side dish of humble crow was well worth it. The akashiyaki was surprisingly light, and the octopus inside was delicious.

No one grows up in the region around Osaka without eating takoyaki, according to Mana Kumagai, president of the Japan Konamon Association. *Konamon* is a Japanese word for foods made with flour dissolved in water. At sixty-one years old, Mana Kumagai represented various companies associated with takoyaki, and she traveled frequently to Europe and Latin America promoting Japanese food and takoyaki in particular. "I was able to prepare takoyaki by the time I was seven years old. It's a part of Osaka's culture. Every house has its own kit for preparing takoyaki," she told me over a plate of eight at the Kumagoro Company restaurant, takoyaki specialists, in the Osaka train station.

As she grew older, Mana Kumagai grew more and more interested in how konamon had found its way into Japanese kitchens. What she learned was that Tomekichi Endo's creation of takoyaki did not come out of the blue but had centuries of backstory. The original konamon dishes are generally thought to have been introduced to Japan from China in the eighth century. They became more popular in the sixteenth century when the Japanese Buddhist monk Sen no Rikyū, who is thought to have been the originator of chanoyu, the tea ceremony, incorporated into his chanoyu practice a crepe-like snack made with flour batter thinly spread on a hot griddle. The tea ceremony, with its all-consuming attention to detail, requires a little something in the way of a snack to accompany the tea, perhaps a precursor to takoyaki.

Gradually, the idea of pouring a flour batter onto a cast-iron pan became part of common Japanese cuisine, particularly in the Osaka region, and by the twentieth century various ingredients were being incorporated. Two basic dishes were created, both of which are popular street foods today. One was *okonomiyaki*, a tasty Japanese pancake made from a batter of flour, egg, and grated yam, and topped with shredded cabbage, sometimes incorporating meat or seafood under a sweet sauce. The second basic batter dish was takoyaki. It was only logical that a dish would be built around madako, since

Osaka borders the Seto Inland Sea, and in those days *O. sinensis* was plentiful and cheap. Takoyaki is often referred to as the "soul food" of Osaka and has spread across Japan. The Osaka version is a little creamier than the crispier Tokyo style, but it is all takoyaki and accounts for a large amount of the annual Japanese consumption of octopus.

Mana Kumagai wrote her university graduate thesis on konamon, and she later founded the association. She turned her thesis into a book, *Takoyaki*, which was published in 1993 and is still available online in Japanese. "Recently, the Japanese have become more interested in learning about the foods we eat, but I was the first one to research and write about the history of takoyaki. No one had done that before," she told me proudly.

For centuries following the introduction of konamon into Japan, wheat flour was an expensive item, found only in the kitchens of the wealthy. Kumagai's research led her to the conclusion that this changed in the mid-nineteenth century when wheat flour–based batter became affordable for the general public. By the mid-twentieth century, takoyaki was being sold as street food to the working class. In Osaka alone there were more than six hundred stalls and restaurants selling takoyaki in 2024, Mana Kumagai said, and of course everyone had their favorites. She offered to show us hers.

We left the restaurant in the train station, each of us eight takoyaki to the good, and followed her to one of the oldest markets in Osaka, which occupies what is said to be the longest covered arcade in Japan. That's saying something given the presence of a covered shopping arcade in just about every Japanese city. The Osaka arcade is named Tenjinbashi Suji, and it is a mile and a half long. Fears are often raised online that Japan's multitude of covered arcades will disappear, driven into extinction by the nation's many vast commercial centers and shopping malls, but there was no sign of that danger as we walked through Tenjinbashi. It was packed with people. We threaded our way through the crowds until Mana brought us to her favorite takoyaki stand, with a name that translated as Lots of Pleasure. A large metal octopus gleamed on the wall above it, and a dozen people waited in line to place their orders with the man behind the griddle. The length of the wait would depend on how many takoyaki the people in front of you in the line had ordered. The price list on the wall started at ¥520 ($3.50) for eight and went up to ¥2,600 ($13) for forty.

Kita Taizo, a thin, intense forty-five-year-old man, was furiously turning takoyaki in their molds, hand moving back and forth across the griddle with his awl, revolving them rapidly and putting them into a cardboard container when they were done, which he passed on to his wife, who added the toppings. His arms and hands were in constant motion over the griddle, but he had a big smile of welcome for Mana Kumagai. As he worked, he told us that his grandfather began this stall in 1953 and his son would carry it on. Takoyaki has supported four generations of his family. "The hardest thing about this work is the temperature," he said. "When it's really hot or cold outside it can be hard to control the temperature, but good takoyaki depends on it."

The other difficult thing about the job was the rising price of octopus, he allowed, as he ladled batter into molds for our takoyaki and dropped a bit of octopus in the middle of each. In fact, Mana Kumagai said her figures showed that the price of boiled octopus from Morocco or Mauritania, which was almost all the octopus used in Osaka's takoyaki, had more than tripled since 1994. Back then, the owner of a takoyaki stall paid ¥128 ($0.85 cents) for one hundred grams of imported octopus, enough to make about sixteen takoyaki. That price had risen to ¥458 ($3.05) by 2024.

We took our takoyaki and sat inside at one of five tables in a small space at the rear of the stall. A little television mounted on the wall displayed sumo wrestling. It was the height of the sumo season, Ian told me, and during his decades living in Japan he had become an avid follower. The takoyaki was delicious. For his eat-in customers, Kita Taizo served it without any sauce on top, but each table had a bowlful and a small paintbrush with which to apply it. A bit of red ginger was served on the side. Many people in Osaka have takoyaki parties, called *takopa*, Mana Kumagai told us. "Students like to do it to celebrate at their school, the boss may bring it to share at the workplace, or families invite other families to their homes. People relax and become more social when they're eating takoyaki, and they enjoy socializing more. I think takoyaki will spread like sushi did."

In the US alone, sushi restaurants reported more than $22 billion in annual revenue in recent years.[8] Even if takoyaki does not become a billion-dollar North American food fad or spring up all over the EU to enhance the already considerable pleasure that Europeans take in socializing, it is likely to soon become better known outside of Japan, and this will only increase the large potential for profits from farmed octopuses. As the octopus supply dwindles, and the market price rises, aquaculture becomes an ever more attractive endeavor. In the past, Japanese landings from the wild varied from year to year, but now no year—good or bad—provides enough octopus for all the Japanese consumers who want it, never mind the rest of the world. In addition to domestic consumption, some of the octopus that Japan imports from other countries will be turned around and exported by Japanese companies, sold at a profit to the growing markets in Europe and the United

States. The octopus consumed in North American sushi restaurants is most likely to have been processed in Japan for shipment to restaurants around the world.

· · · · · · · ·

From Osaka, we crossed the country by train back to the east coast and Sendai, the hilly, coastal city of a million people where Ian Gleadall lived. The day after we arrived in Sendai, Ian picked me up at the hotel in his car and we headed north. While the city is hilly, it is built on a flat coastal plain, and when we left it behind the highway passed beside miles and miles of rice paddies. We were on our way to Ishinomaki, north of Sendai, a city of about 150,000 people, which had been devastated by the March 11, 2011, tsunami. Ian wanted to show me an octopus research station he had overseen, built on a bluff overlooking the sea not long after the area was struck by the tsunami. It still had a few tanks with a handful of octopuses in them, but nothing much was currently happening there, he said.

Nationally, interest in octopus farming had grown exponentially after the tsunami along Japan's east coast. The tsunami took nearly twenty thousand lives, erasing a whole swath of the coast, destroying whole neighborhoods, villages, and towns. In Ishinomaki we saw block after block with brand-new buildings and houses, everything rebuilt since 2011. The nuclear plant at Fukushima, some one hundred miles south of Ishinomaki, was damaged in the second-worst nuclear plant disaster in history, behind only the 1986 meltdown at Chernobyl. Life-threatening quantities of radiation were released, and the area within a 12.5-mile radius of Fukushima was labeled an exclusion zone. People living within it were subjected to a mandatory evacuation, and fishing was prohibited in the part of the exclusion zone extending into the Pacific Ocean. Overnight, Japan's large east coast octopus fishery was shut down as radioactive water from the plant's cooling system leaked into the ocean.

A fisherman in Ishinomaki told Ian Gleadall that he and his fellow fishermen had taken their boats out to sea in order to ride out the impending earthquake and possible tidal wave. When the danger passed, he came back to port, his boat safe and sound, only to find that his daughter's school had been in the tsunami's path and all the students, including his daughter, had drowned while he was out at sea. These were the sorts of horrific memories shared by many, many people who lived through the tsunami, Ian told me as we drove through the rebuilt city.

It was not until July 24, 2012, that octopuses from those coastal waters were approved to reappear in markets. They carried a certification from the Fukushima Prefectural Fisherman's Association that they were free of radioactivity. "The octopuses . . . are sweet and tasty. Products from Fukushima are finally back on the market, and we're going to do our best to sell them as part of the reconstruction effort," Takashi Suzuki, a local seafood retailer, told the *Japan Times*.[9]

Ian Gleadall was working out of the country when the earthquake struck, but his wife and children were in Sendai, in the middle of one of the hardest hit areas. Their home was not damaged, but his wife, Etsuko, spent the next months going to one of the most devastated areas of the prefecture to help with relief efforts. And even though he was away when it happened, the earthquake and tsunami had an impact on Gleadall's professional life.

From the outset of his university career, he had been reluctant to do research into farming octopus. Octopus aquaculture seemed like a complicated, difficult, long-term undertaking to him, and he was not particularly interested in working on it. He was quite happy doing octopus taxonomy, thank you—and very good at it. When an unusual or unidentified octopus was found along the coast anywhere near Sendai, Ian was likely to get a call to come and confirm an identification. For him, the height of professional pleasure was the chance to visit the archives and basement collections of museums around the world, places like the British Museum or the Smithsonian, and search out old specimens of octopuses on dusty shelves or forgotten in drawers, read their labels to determine if they had been taxonomically classified, and, if so, determine whether the classification was correct.

"Nowadays many researchers think that if a genetic sample is taken it's enough to use for identification," he told me. "But I believe there's no sense in just having a tissue sample and determining DNA; you need the animal to go with it, you need to *see* the whole animal, and that kind of description, combined with genetic information, is what you need to do a good job."

The 2011 tsunami, and the subsequent shutdown of the fishery along the Pacific Ocean coast of Japan, was a wake-up call for the Japanese government. It was a stark example of the fragility of the nation's marine resources, and the authorities began to see aquaculture as an invaluable part of guarding against the disappearance of species like *O. sinensis* from Japan's tables. Suddenly, funding for octopus aquaculture appeared.

Despite his reluctance, following the tsunami Ian Gleadall was basically press-ganged along with a number of other marine biologists into octopus aquaculture research. The funds were available, and his university committed to participating in the resulting wave of investigation. Numerous fish and shellfish were already being farmed in Japan; every year aquaculture's share of the market was growing. It appeared that if the Japanese wanted to continue eating large quantities of octopuses at reasonable prices, aquaculture would be indispensable.

"I didn't really choose to do octopus aquaculture," he told me. "I'm a taxonomist, and aquaculture is not my field. I liked doing taxonomy because I didn't need much money to do it, and I didn't need live animals. But after the tsunami disaster, this company named Hotland [the parent company of Gindaco] came to find someone who knew specifically about octopuses and sign them up to do octopus aquaculture."

In 2014, Ian Gleadall became part of a group of researchers in Japan who formed the joint academic-corporate coalition called the Octopus Aquaculture Development Project, initiated by Hotland, with the goal of farming octopuses in land-based tanks. They were an eclectic group of academics, industrialists, and businesspeople, including representatives from Hotland and three universities. They described themselves in an online statement as a collection of "determined, like-minded individuals . . . with one common aim: to successfully culture the common Asian octopus on a commercially sound basis. . . . Various strategies towards attacking the problems of raising paralarvae are underway at 4 different locations."[10]

However, try as they might, it has not happened. They were able to raise octopuses to an edible size if they started with juveniles, but the difficulties of keeping paralarvae alive until they grew into juveniles continued to block their progress. Many of the group's members gradually devoted less time to the project and turned their attention to other things. "The project is still ongoing, and there are still businesses and three universities involved," Ian told me. "It's hard to say what happened, but I think the project has always suffered from a lack of large numbers of people working full-time on octopus aquaculture. Most of the people have also been involved in other projects of their own, so a full focus on octopus aquaculture has never really happened, except for when I was working on it.

"Before I retired, I had only one full-time technician and some undergraduate students. We were trying to figure out what happens when the common octopus paralarvae come down to the sea floor, and what they might be eating. They are only about two centimeters long, and we don't know much really about what kind of life they lead once they come down to the seabed. Things like what they do, what they eat, how they escape predators. It's just a blank. If it were up to me, I'd find a good site, set up a large number of tanks for experiments, and employ a large number of researchers to work on it. It needs heavy focus, no doubt about it, and with just a few researchers working on it like myself, there's a limit to what can be achieved.

"It's still an important research effort in Japan. There's a continual increase in demand for octopus products and at the same time a decrease in the amount that's caught. The last three years in particular have been dire. So the government is still willing to fund octopus research. The money our project is getting comes from the Japan Science and Technology Agency, and they're more interested in projects that combine university researchers with [private] companies. There are not many like that in Japan."

Ian was retired from the university but still part of the Octopus Aquaculture Development Project and trying to grow *O. sinensis*. The project's research contract with Hotland was in force until 2030, but it wasn't nearly enough money to pay for sufficient personnel or to fully equip a laboratory. Another researcher who was brought into the project by Ian Gleadall was Masazumi Nishikawa, a professor of food science and technology at Miyagi University in Sendai. Like his colleagues', his enthusiasm for the project had waned over the years.

"I have been working on developing a feed for the paralarvae," he told us. "I am putting it in tiny four-hundred-micron capsules, and they are eating them, but not in sufficient quantities for real growth, so far. Our project is set to run through 2030, when it is supposed to be producing octopus at an industrial level. So in the next couple of years we've got to get to the stage where we can then develop it for industrial production. That means in the lab over the next couple of years we've got to get the basic stuff right to where we can do it in the lab and move on from there."

Ian Gleadall concurred with his colleague, nodding with an air of resignation. Nishikawa continued, "One of the problems after they've settled on the seabed is what to feed them after that. That's when there also tends to be a lot of cannibalism. When they first settle it's okay, but after about a week or so that's when the cannibalism sets in, so the problem is what to feed them. That's something that has to be developed. If we solve the problem of what to feed them it should be possible to succeed in farming them. It's a difficult problem to solve, but I'm confident if we had the resources we could do it by finding the right feed and [determining] how to keep them separated in tanks to avoid cannibalism."

Masazumi Nishikawa grew up in a village on the Japan Sea coast, and many of his family meals included octopus in a stew with potatoes or boiled, sliced thin, and served with rice. As a youngster, he told me, he spent a lot of time by the sea, and when the time came to elect a specialty at the university, marine biology seemed like a natural choice. He spent the first twenty years of his career developing feeds for fish aquaculture, working for one of Japan's largest seafood companies, Maruha Nichiro, and the last twenty years on the faculty at Miyagi University.

The consortium of business and academic interests in which Ian Gleadall and Masazumi Nishikawa were participating was a rarity in Japan, Ian told me. Most Japanese corporations were not amenable to collaborating with a public entity. Hotland was an exception in its cooperation with the university researchers, while others had decided to try and fill the national need for octopus aquaculture with research at publicly owned universities or with wholly private efforts using their own labs and staffing them with full-time researchers.

One such company was the Japanese seafood giant Nippon Suisan Kaisha. Better known as Nissui, it had a huge presence in the Japanese frozen seafood market. In June 2017, Nissui announced that its researchers had successfully closed the reproductive cycle of *O. sinensis* at its laboratory in the Oita Marine Biological Technology Center beside the Seto Inland Sea.[11] A Nissui spokesperson told reporters that researchers there had raised adult octopuses from 140,000 eggs of an earlier generation of aquacultured animals.

The company declared that further research was needed to ensure a survival rate sufficient for commercial success but predicted that it would put farm-raised octopuses on the market by 2020. There seemed no reason to doubt it. Nissui was already heavily invested in fish aquaculture, reporting almost $500 million in sales of its farmed fish; coho salmon and trout from Chile made up 62 percent of its aquaculture production.[12] Inside Japan, the company farmed Japanese amberjack, bluefin tuna, coho salmon, mackerel, and sea bream, mostly in net pens in the ocean. But 2020 came and went without a single farmed octopus sent to market, and by 2025 that was still the case.

Nissui's prediction of farmed octopus by 2020 turned out to be as overly optimistic as all the other predictions that have been made around the world about imminent success in farming octopus. Nissui is even more secretive about its plans than Galicia's Nueva Pescanova, and no one knows how much progress, if any, has been made since that 2017 announcement. In late 2024, a spokesperson for Nissui succinctly answered my email query: "Our octopus farming is still in the research stage, and there are no plans to ship it for the time being." When I asked permission to visit the lab and speak with researchers, my request was summarily rebuffed.

"I think Nissui is in the same boat as everybody else," Ian Gleadall told me. "I don't hear much [about them]. I know one or two people who know one or two people who supposedly work there, but they keep their information very close; there's not much coming out. Frankly, for the last couple of years I haven't heard anything from there."

The Japanese government was funding octopus aquaculture investigation at two research centers—one in Hiroshima and one on Shikoku Island—but they were no more forthcoming than Nissui despite the public funds involved. Researchers at both centers declined Ian's requests on my behalf for visits or interviews. Gindaco, the huge takoyaki franchiser, and its parent company, Hotland, have funded research into octopus farming but did not respond to repeated requests for an interview.

An exception to the secrecy surrounding research was in the laboratory of a marine biologist named Shigeki Dan, a friend of Ian Gleadall's and an associate professor of marine bioscience at the Tokyo University of Marine Science and Technology. We took a train from Sendai down to Tokyo one morning. The university, on the outskirts of Tokyo, was a little island of quiet tucked into a busy urban neighborhood. Fallen golden leaves of ginkgo trees carpeted the walkway to Shigeki Dan's laboratory. One building we walked past had a large picture window, behind which a whale skeleton was suspended from the ceiling.

Shigeki Dan was leading a team affiliated with the research center in Hiroshima. His group had a major breakthrough in 2021 when they developed a system of aeration and water flow that would keep paralarvae on the bottom of a tank, rather than passively drifting.[13] With this technique, the Tokyo team had achieved an unprecedented and astonishing paralarval survival rate of 90 percent.

Shigeki Dan welcomed us into his office. He was a short, squarely built man, happy to talk about his work, still excited by the research. A tall cabinet held a shelf of glass bottles, the size of pill bottles, each holding octopus eggs or a tiny octopus in formalin. On the top of the cabinet were a half-dozen small ceramic octopus pots pulled up in the nets of various trawlers, which he said came from the Seto Inland Sea and were more than fifteen hundred years old. Until 2015, he was working with crab larvae, and he became involved in octopus aquaculture when crab larvae were being investigated as a possible feed for baby octopuses. He laughed. "I saw it was easier to find funds for working with octopus than with crabs."

Shigeki Dan's researchers were housed in a low two-story concrete building, where labs and classrooms occupied a large wing of the second floor. He led us through a room with more than a dozen students and researchers at work on their computers, then on to a room holding a tank with his aeration system at work. A few paralarvae, looking no bigger than specks of dust, floated in the water while a mass of them were huddled close together at the bottom of the container. This tank represented the cutting edge of Japanese octopus aquaculture research. "Our survival rate is wonderful, but the problem is that then they become weakened as juveniles and die. I think we can solve this, but I think it will be another ten years before we are able to say we can farm octopus. I'm forty-six, so I have another twenty years to try and grow them," he laughed some more.

"It will only become more and more important in Japan to aquaculture octopus. They should be taking about ten thousand tons a year from the Seto Inland Sea, and now that's down to around two thousand. I am from Shikoku Island, and for us it was a common food and used to be really cheap. That's why when I was a child, my grandfather used to make a big pot of octopus stew. It was cheap. You used to be able to buy thirty takoyaki for three hundred yen [$2], and now for eight takoyaki you have to pay eight hundred yen[$5.50].

"There are many reasons for the decline in the number of octopus being caught. One reason is that the currents are changing, and a very warm current relatively low in nutrients is flowing into the Seto Inland Sea, which is relatively high in nutrients. This did not used to happen, and the waters are becoming less rich. An octopus can produce as many as two hundred thousand eggs a year. If there is plenty of food to eat as they're growing, it's possible to keep catching ten thousand tons a year. That was happening in the 1990s and 2000s. I believe the problem is not overfishing—it's that they don't have enough to eat."

Once the problem of getting the paralarvae to settle on the bottom was solved and they could grow into juveniles, it was not a far reach to think that shortly Shigeki Dan's team would be growing them into marketable adults. But his team had run straight into the same puzzle that would-be octopus farmers around the world were still trying to work out: what to feed them. Live prey, far and away the favorite meal of octopuses, is too expensive, way too expensive. There had been a moment's enthusiasm for the possibility of using crab larvae as feed for the smallest octopus, but it was too complicated and difficult to do correctly. Shigeki Dan told us he had concluded, at least for the moment, that the key lay with artemia, and he showed us another room in which a tank the size of a large home aquarium held a swarm of swimming artemia. While these tiny sea monkeys alone will not produce all the protein required by growing octopuses, he was experimenting with supplements to the artemia's feed that would transform them into a complete nutritional package for octopuses. What those supplements were, he said with an apologetic laugh, he was not going to tell us. "We are working on finding the right supplement. That's our secret, and that's my dream: to be able to produce Japanese octopuses using artemia."

With thanks and many a bow, we bid Shigeki Dan goodbye and took a subway back to Tokyo's center. That evening, Ian Gleadall insisted on bringing me to one of his favorite watering holes in Japan, the Rising Sun, the oldest English pub in Tokyo, which had recently celebrated the fiftieth anniversary of its founding. Whenever he found himself in Tokyo, Ian passed time in the pub, he told me, and he even took an occasional train ride down from Sendai for no other reason than to visit it. Over pints and plates of shepherd's pie, he did not try to hide the admiration that our visit to Shigeki Dan had generated. "Did you see all that equipment and all those students and researchers? That really makes me envious. That's fantastic what's going on there. And artemia! That's really surprising. He's using artemia. I'm amazed."

Despite a decade spent doing research in farming octopus, Ian admitted he was still far from certain that the cephalopods were viable candidates for commercial farming, at least in the near future. The problems of what to feed them and how to keep them alive remained vexing questions. The researchers with whom he was affiliated in the Hotland group had failed over the past decade to make much progress.

He was following Nueva Pescanova's efforts from afar and had visited the Octopus Project in Sisal, but he told me he remained unconvinced that either of them would put a product on the market anytime soon. He did not see them as competition but as colleagues in a frustrating and difficult research effort. He told me that he had found at least twenty places in the world where somebody was working on octopus aquaculture. "There are several places in South America, in South Africa, various places in Europe, Australia, and other parts of Asia, so here and there a lot is going on, really. I don't see any of them as competition. Most of the aquaculture people just want to help each other because it's so difficult."

· · · · · · · ·

The Japanese taste for food from the ocean has long been the target of animal rights groups around the world. For example, documentary films and books assailing the taste for whale, and the market it creates, has curtailed the national consumption of cetaceans but not eliminated it. Pride in the nation's history and culture is strong, and the Japanese tend to resist when pressured by outsiders to change their ways. Researchers in Japan were acutely aware of the push in the United States for legislation banning octopus farming. Ian Gleadall and a number of his colleagues sent an open letter to *Science* magazine defending the potential for octopus farming. *Science* rejected it, as did other journals. It seemed to Ian that no one wanted to hear about why it might be a good idea to farm octopuses.

The animal occupies an outsize place in Japanese culture, reaching far beyond its prominent role in the national cuisine. For instance, octopuses often figure in erotic Japanese art and literature and have done so for centuries. The nineteenth century saw a proliferation of *shunga*, a Japanese school of erotic prints and paintings, which were sophisticated and remarkable for both style and subject matter.

The Japanese artist Katsushika Hokusai was one of the best. He is known for his woodblock series *Thirty-Six Views of Mt. Fuji* as well as the oft-reproduced *The Great Wave Off Kanagawa*, which depicts a huge wave looming over a tiny boat. Another famous colored woodblock print by Hokusai is the astonishing *Dream of the Fisherman's Wife*, in which a naked, young, reclining Japanese woman is sexually pleasured by two octopuses. The scene has its roots in a thousand-year-old legend, *Taishokan*, which tells of an octopus-like monster and a girl fighting over a jewel.

A lot of shunga, including prints of *Dream of the Fisherman's Wife*, made their way to a more buttoned-up Europe during the nineteenth century and had a strong influence on Western artists like Pablo Picasso, Gustav Klimt, and Auguste Rodin, all of whom had collections of erotic prints from Japan.[14] Pablo Picasso told Brassaï why he so admired Hokusai: "He revealed a new way of conceiving art, freer and more alive, more modern and original."[15]

As with so many other cultural icons in the digital age, the elegance and thrill of Hokusai's print, and of shunga in general, has been reduced and cheapened from art to something that serves a commercial purpose. The exquisite work of those nineteenth-century shunga artists has devolved into *hentai*, Japanese internet pornography featuring octopuses, tentacles, and young girls. Tens of thousands of hentai videos have been produced, initially fueled by the Japanese prohibition on showing actual genitalia and intercourse. The male organ is often represented by a tentacle sliding over a young woman's body.

In addition to being present in the pornography market, the octopus is treated with reverence as in the Temple of Jojuin in Tokyo, where the animal is exalted. Dedicated in 858 AD by the Buddhist monk Ennin, the temple is rooted in a legend about an octopus saving Ennin's small statue of Yakushi, the Buddha of healing, when it fell overboard while the monk was on a ship from China to Japan. The temple is called Takoyakushi and is decorated with paintings and sculptures of Buddhas and octopuses. Kyoto has its own Takoyakushi temple, built in 1181 and dedicated to the Buddha of medicine. It is believed to contain the power to heal, and visitors come to rub their left hand on a wooden octopus and ask to be cured of their ills.

In a crowded Sendai neighborhood, tucked away behind offices and parking garages in a quiet space of a full city block, this city has its own Takoyakushi shrine with images of octopuses on wooden posts and ancient stone Buddha statues bordering the grounds. Behind a bustling business center, it took Ian and I a half hour of searching to find it, even with Google Maps in hand. There is said to have been a shrine on this spot since the twelfth century. This particular one is known for its power to cure warts. It was autumn, and Sendai's sidewalks were planted in rows of ginkgo trees, stretching to the horizon, their autumn leaves like beacons of golden light in the distance. A short elderly man at the shrine wielded a straw broom, sweeping up the ginkgo leaves scattered around the graveled courtyard. It was a quiet and peaceful space, with benches where people could sit and take in a bit of stillness.

An article in a Buddhist journal described the octopus's attraction: "Their [octopuses] very appearance is disconcerting without appearing (unless there is artistic exaggeration) threatening, and can even look cute and endearing, with tako characters being a fixture in Japanese pop culture and anime. The mystery and mystique of cephalopods perhaps captures, visually and artistically, the inexpressible, ineffable nature of enlightened beings."[16]

TRADE ROUTES

FOR ALL the various roles that octopuses play in Japanese culture, its most prominent role is on a plate, and the taste for octopus makes Japan an important player in the global market. The concept of a worldwide octopus market did not even exist until the second half of the twentieth century. Before that, no one would have thought of farming them because demand was exceedingly limited. The vast majority of people in Kansas City, Saint Petersburg, Prague, London, or Paris lived their whole lives without ever seeing octopus on a plate. They would likely have laughed at the thought of eating one. Even for those who had octopus available close to hand, its value was small or nonexistent. Octopus was something eaten only in those countries where it was caught, and more than enough were taken from Japanese waters to satisfy national demand.

That had changed drastically by 2025, to the point that my hopes of being able to buy a takoyaki griddle in Barcelona were close to becoming a reality. A fifteen-minute walk from my flat would now take me to Takopa, one of Barcelona's three restaurants specializing in takoyaki, two of which were owned by Quim Rueda Morales, a thirty-two-year-old Catalonian, and his investors. Quim spent two years living in Japan studying traditional Japanese art and calligraphy, working at whatever he could to get by.

"The first time I tried takoyaki was in Kyoto," he told me. "The cherry trees were in bloom. It was really emotional. I was celebrating my birthday in a bar and the owner made takoyaki. It was a revelation. But as much as I liked it, I had no idea that I would eventually have a business dedicated to takoyaki."

When he returned to Barcelona, he began to hone his takoyaki skills. "I started researching how to treat octopus in the kitchen, how to prepare takoyaki so as to be sure the octopus releases its juices, have the color the correct pink to mix with the dough, make sure of the exact temperature of the griddle."

When he opened Takopa's doors in 2021, it was strictly for takeout, but as time went on, word spread, and he put some tables in the restaurant, then opened up a second location. He charged a reasonable six euros for six takoyaki balls. He predicts a growing presence of takoyaki shops in European cities and shares Mana Kumagai's enthusiasm for takoyaki as an excellent facilitator of sociability. "I've got a griddle in the house, and sometimes I'll make takoyaki at home with my wife, and other times we'll take it to my parents' home with my sister. The takoyaki is really the excuse to get together with your family. It's not just eating takoyaki, it's the takopa, the party. What's more, in Spain, the culture of the tapa is very ingrained, and a plate with six takoyaki is very similar to a tapa—a small plate of food that's tasty, easy to eat, and goes very well with a beer. It fits perfectly with our culture of enjoying food together with friends."

Quim's ingredients were sourced locally. The octopus came from the morning's auction of freshly caught fish and seafood in Barcelona's wholesale market. He believes this puts his takoyaki a cut above what you would find in other places around the world, including Japan, where the great majority of octopus used in takoyaki is not caught locally. Between 1950 and 2014, global octopus landings increased eightfold. Much of the catch went to satisfy Asian demand as landings from local waters decreased. An important octopus market exists in Mediterranean countries like Spain, Italy, and Greece, but the Asian market for octopus is, indisputably, the driving force behind the worldwide trade.

By far the largest part of the octopus imported by Asian countries comes from the waters off North Africa. The Saharan Bank, situated off the northwest African coast, is considered one of the richest fishing grounds in the world, blessed with a broad continental shelf and a large, permanent upwelling where nutrient-rich deeper waters rise toward the surface. The area has supported an artisanal fishery for centuries but has never come under the kind of industrial pressure it is currently experiencing.

Japan and South Korea are huge octopus consumers. South Korea accounted for $128 million of octopus imports in 2021; Japan and China were not far behind.[1] China is, by far, the world's largest exporter of octopus, and the octopus fishing fleets of Chinese companies ply waters around the globe, from North Africa to Chile. The Japanese export a good deal of octopus, though much more is imported from North Africa and stays in the country to satisfy the Japanese appetite. That is on top of what is caught locally. Japan's domestic catch has fluctuated in recent decades between forty thousand and sixty thousand metric tons but is currently in decline.[2] A great deal of the octopus eaten in Japan lived its life in foreign waters.

The arduous journey made by the Galician arrieros in their mule carts with dried octopuses for sale, traveling from Spain's northwestern coast to remote inland villages, was the eighteenth-century version of today's global octopus network—big refrigerated freighters moving shiploads of frozen octopus around the world in a commerce worth some $2.7 billion a year. The journey from Morocco, Mauritania, or Senegal to Asia can take up to two months, during which the frozen octopus has to be held in ideal conditions.

In the first three months of 2023, Japan imported 10,434 tons of octopus, and 38 percent of that came from Mauritania, according to the United Nations's Food and Agriculture Organization. In 2021, octopus exports from Mauritania were worth about $312 million. Most of the octopus that Japan imports from Mauritania comes from Nouadhibou, the nation's second-largest city and its largest port.[3]

Mauritania's huge octopus fishery is the result of one Japanese man's efforts to expand the flow of octopus to Japan and to help the local Mauritanian fishers. In the mid-1970s, Masaaki Nakamura was twenty-nine years old, had a PhD in fisheries science, and worked for the Japan International Cooperation Agency. They sent him to Mauritania to see whether there was anything of commercial interest in the country. Nakamura told the *Daily Maverick* in a 2023 interview, "There was nothing. Nothing but desert. I didn't know what I should do."[4]

He wound up in Nouadhibou, and there he found something other than desert. He found the Saharan Bank and octopuses. Lots of octopuses. The Japanese were much more demanding than Europeans about the octopus they bought and ate. As long as there were enough to meet demand from Japanese waters, they were uninterested in importing them. But Masaaki

Nakamura foresaw a time when the Japanese would need to augment their local landings. He hired a handful of Mauritanian fishermen and taught them how to land and handle octopus that would be saleable on the Japanese market. It was a process he began by showing them how they could fish for octopus with ceramic pots so they would not have to buy bait, followed by how to carefully handle and process the octopus they caught. He found a Japanese company willing to buy the Mauritanian product and initiated the first exports to Japan—a trickle that would first grow into a stream, and then a river.

Almost fifty years later, Masaaki Nakamura was well into his seventies and still shuttling between Tokyo and Nouadhibou, watching over what had grown into an important commerce over the decades—too important in some ways, he said in his 2023 interview with the *Daily Maverick*. In 1993, a reported four hundred pirogues from Nouadhibou were fishing for octopus. By 2023, there was a fleet of about seventy-five hundred pirogues crowded into the port. They were crewed by an estimated fifty thousand artisanal fishers. One hundred thousand people were dependent on the octopus fishery for their livelihoods, with about 80 percent of Nouadhibou's available employment related to octopus fishing.

In this city of 140,000 inhabitants there were fifty plants to process octopus, exporting principally to Japan, China, and Spain. Masaaki Nakamura's original idea of creating an octopus fishery for export to Japan grew out of his conviction that in addition to providing a reliable source of octopus to the Japanese, it would benefit a large number of poor Mauritanian artisanal fishers and their families. He watched with growing alarm as refrigerated Chinese trawlers began to capture thousands of tons of octopus to sell to Japan and South Korea. He did not mince words. "China has a huge fleet of hundreds of trawlers off West Africa," he told the *Daily Maverick*, "and they are stealing all the resources."

Li Yemin is a sales manager in Nouadhibou for the Chinese seafood giant Hongdong International Fishery Development Company. She was born and raised in China and studied in Spain. Her employer is the largest Chinese company fishing for octopus out of Nouadhibou. We met at the Hongdong booth, set back in a corner of the annual Seafood Expo in Barcelona's huge convention hall. Li Yemin was in her early thirties; both her English and her Spanish were impeccable. The expo's congress drew over thirty-five thousand marine products professionals in 2024. Among the hundreds of booths offering information about fish and seafood, only a few represented octopus vendors. Not many of the expo's attendees were there with octopus on their minds, but Li Yemin was ready for those who were, with large photos of octopuses and freezer ships decorating the walls of the booth. Hongdong had fifteen trawlers fishing the waters off the coasts of Western Sahara and Mauritania for octopus, she told me, and it was the company's bestselling product.

Hongdong's fleet of trawlers captured and froze some 4.5 million pounds of octopus a year, which were taken to the company's factory in the Canary Islands. There, they were processed and shipped to China, where they were eaten or packaged and exported once again. Li Yemen complained that her job was getting more difficult with the numbers of octopuses landed in Mauritanian waters in the past two years going steadily down, in addition to the Mauritanian government deciding every year in what zone the boats could trawl, which could vary from one year to the next. She told me in a resigned voice that keeping up with changing regulations was a stressful part of her job and kept her busy, but when it came time to relax, living and working in Nouadhibou did not offer much in the way of opportunities. She was happy for the chance to spend a few days in Barcelona. "All I've done for two and a half years in Nouadhibou is eat, sleep, and work."

Shoppers in Mauritania will not find Hongdong's octopus on supermarket shelves. It all goes to China, and even the wholesale price of Mauritanian octopus puts it far beyond the reach of the country's average food shopper. Mauritania is not only a poor country but an oppressive one. It was the last nation in the world to abolish slavery—in 1981—and despite the professed prohibition, the nonprofit Global Slavery Index estimates that in 2021 some 149,000 people were slaves, living in forced labor or forced marriage, many of the laborers Black Mauritanians enslaved by Arabs.[5] Despite a plentitude of resources like oil, iron, gold, octopus, and fish, in 2024 over 28 percent of the population lived below the international poverty line of $3.65 per day, according to World Bank figures.[6]

As early as the mid-1970s, the Japanese had begun importing octopuses caught in the Atlantic off the coasts of Spain and Portugal, but these stocks did not take too long to exhaust. Fishing fleets moved south. The North African nations knew a good thing when they saw it and charged millions of euros to foreign trawler fleets for fishing rights.[7] In 2012, under pressure from its artisanal sector, the government ruled that only Mauritanian boats could fish for octopus within a two-hundred-mile exclusion zone off its coast. Even so, landings continued to decrease. Under a six-year agreement signed in 2021, the European Union paid €57.5 million a year, plus a one-time payment of €16.5 million, so that EU-flagged vessels could fish in Mauritania's exclusion zone.[8]

The octopus market generated over $300 million in Mauritania in 2021, and government officials exercised strict control over the zones that could be fished and who could fish them. Octopus fishing during the reproductive season was strictly prohibited as they tried not to kill the goose laying all those golden eggs. Nevertheless, the *O. vulgaris* population continued to decline, and competition among the fishers and the companies they dealt with grew ever fiercer.

Babana Yayha Emhamed, an inspector general with the Mauritanian Fisheries and Maritime Economy Ministry, told the *Daily Maverick* in 2023, "The fish eat each other, and the fishermen do the same. There's the big importer who buys the product and tries to manipulate Mauritanian producers. These manipulate the middlemen, and they in turn manipulate the artisanal fishers. Each is the slave of the other."[9]

When Masaaki Nakamura began buying and sending octopus to Japan, the cephalopods were still plentiful off the coasts of Portugal and Spain; these countries were where the big fishing trawlers worked. Nevertheless, the new North African exports found a ready market, which only grew larger over the years as the octopus population declined off the European coast. Nouadhibou is only about forty miles south of Mauritania's border with the Western Sahara, and it didn't take long for the artisanal fishers of the small villages along that coast to learn that there was a market for octopus. Japanese seafood dealers began expanding their buying activities north from Mauritania. The Japanese presence in North Africa has been significant both economically and politically. Mauritania was where the Japanese landed first, thanks to Masaaki Nakamura, but as early as the mid-1980s Japanese companies began moving north from Mauritania and establishing themselves in the part of the Western Sahara claimed by Morocco. One of the first places north of Mauritania to begin a large-scale exploitation of the Saharan Bank cephalopod resource was Dakhla, on the coast of the Western Sahara.

The artisanal fishers of Dakhla, about 230 miles north of Nouadhibou, began to discover an octopus market far beyond the local one in their hot, sandy town bordered by the ocean and the Sahara Desert. Dakhla is one of the oldest settlements in the Western Sahara, founded as a Spanish colonial outpost in 1502, when it was called Villa Cisneros. It was used as a prison camp by the Spaniards during the Spanish Civil War, but shortly before the dictator Francisco Franco died in 1975, he renounced Spain's territorial claim to the Western Sahara and withdrew all Spanish troops under an agreement for a joint Mauritanian-Moroccan administration.

The Saharwi people assumed that their land would become independent, but it didn't happen. The Moroccans coveted the potash that could be mined there in great quantities and laid claim to it as part of Morocco. Mauritania, not looking for a confrontation with its more powerful neighbor to the north, renounced its part of the deal, sparking a war between Moroccan forces and the Polisario Front, the military arm of the independence movement in the Western Sahara. An entire generation of Sahrawis were evacuated from their homes and sent into exile in refugee camps at Tinduf, in the desert of western Algeria. The war with Morocco lasted until a cease-fire was brokered by the United Nations in 1991, with an agreement for an independence referendum to be held in the near future. Today, almost forty years later, the referendum has yet to take place. The long-lasting conflict has been one of the world's most intractable. A generation has come of age in Tinduf without ever setting foot in their homeland.

In late 2020, the United States, under Donald Trump's administration, did what no previous administration—Republican or Democrat—had been willing to do and officially recognized the Western Sahara as belonging to Morocco. The "Proclamation of Recognition" also pledged to open a US consulate in Dakhla in order "to promote economic and business opportunities for the region."[10] The Biden administration suspended the plan as a Democratic majority in the Senate moved to withhold any funds to be dedicated to building the consulate, but following Trump's victory in the 2024 election, a Polisario spokesman expressed worry that the consulate might soon be opened.[11]

In 2023, Pedro Sanchez, the prime minister of Spain, also recognized Moroccan sovereignty in the Western Sahara, in order to obtain Moroccan cooperation in halting the flood of undocumented migrants to Spain. And in late 2025, the United Nations officially declared in favor of Morocco. The Polisario has vowed to continue the fight for their land, but as support dwindles, it appears increasingly unlikely that Sahrawis will ever be able to return home from exile to an independent Western Sahara.

Japan's appetite for octopus was destined to play a significant role in the struggle over Western Sahara independence. It turned out that in addition to potash, the region held access to another important resource: octopus. What was an informal artisanal fishery for some Sahrawis living in towns like Dakhla along the coast, and a source of food for their family tables, grew into a multimillion-dollar business. The Japanese funded and designed the construction of a fishing industry. The Moroccan government began encouraging its citizens to move south to Dakhla, where there was work fishing for, processing, and shipping octopus, and jobs sprung up in industries like the processing plants or with the businesses that provisioned the octopus boom. Japanese quality control supervisors appeared; they too needed to be housed and fed. Until then, Dakhla's only attraction to foreign visitors was the stiff offshore winds that blow there some 330 days a year, making it one of the best places in the world for wind surfing.

Japan was not the only large importer of *O. vulgaris* from Moroccan waters. South Korean seafood companies followed close behind the Japanese. In 2022, European Union countries imported over forty thousand metric tons of it, almost of all of which went to three countries: Spain, Italy, and Greece, with Spain importing more than twice as much as Italy and five times as much as Greece. That year, the export of Moroccan octopuses to those three European countries was worth almost half a billion dollars.[12] Most of them came from the Western Sahara coast, with 73 percent of Morocco's export of fish and cephalopods coming from the Saharan Bank in 2020.[13]

This has meant a steady growth in Dakhla. For the first decade of the twenty-first century, the influx of Moroccans working in the octopus fishery created slum neighborhoods full of insalubrious shanties. In 2010, one-third of Dakhla's residents lived in substandard housing, but the Moroccan government was encouraging an ever-growing number of Moroccans to settle in Dakhla and fish for octopus. To provide for them, beginning in 2010, the government initiated an aggressive program to build better housing called City Without Slums for those who moved to Dakhla.[14] The result was an emerging constituency who could be counted on to cast their votes in Morocco's favor if the long-promised independence referendum was ever held, a possibility that seemed considerably less likely following recognition of Moroccan sovereignty by the United States and Spain.

"Japanese financial and technical support to Morocco's fishing industry has not only led to unbelievable development and growth—turning it from a lawless, artisanal and unproductive sector, into a high-tech and profitable industry—but the cooperation has also had an indirect effect on internal politics concerning the incorporation of Western Sahara. In a win-win situation, Japan benefits from Morocco's fisheries products: the octopus collected in Dakhla alone constitute 40–60 percent of the Japanese market demand," wrote Mayuka Tanabe, a fellow at the International Institute for Asian Studies, celebrating the cooperation between Morocco and Japan.[15]

Not everyone sees it that way. The Sahrawi who were the original artisanal fishers ("lawless" and "unproductive," according to Mayuka Tanabe) have, naturally, viewed these developments with dismay. Between 1994, when the first octopus freezer unit was built in Dakhla, and 2004, when some ninety companies linked to frozen octopus were registered, the local fishers watched as thousands of Moroccans came from other parts of the country to settle in Dakhla and compete with them for the catch.[16] Numerous violent conflicts have broken out, with Sahrawi octopus fishers clashing with their Moroccan counterparts as well as with Moroccan police and military.

The North African fishery has had to take drastic steps to protect the remaining octopus stocks. In 2000, for instance, when the octopus gold rush was just beginning, Morocco reported landings of 99,400 metric tons of octopus, but only four years later, in 2004, that had dropped to just 19,200 tons. Morocco began to more strictly regulate its octopus fishery, though the numbers have still not fully recovered twenty years later. Still, the country is again the world's largest exporter of octopus, with a total of 50,943 tons in 2020.

In addition, farther offshore, the European and Chinese trawlers have made a significant dent in the octopus population. The intense fishing off the North African coast has meant that more and more octopuses are now landed further south. In the past decade, freezer-equipped octopus fishing vessels worked their way down beyond Dakhla and Nouadhibou to sub-Saharan Senegalese waters. They arrived to find—no surprise—that the Japanese were already there buying Senegal's *O. vulgaris*. As was the case in Mauritania, a single person from a Japanese development agency had begun the organization of an important octopus export route to Japan.

In 2003, some octopuses from Senegalese waters were being sold to European countries but not to Japan. The Senegalese eat a lot of fish, but they do not eat octopus; they therefore had little sense of how to treat it as food. They assumed that as long as an octopus was alive when it was caught, it was edible and fit for the foreign market. The quality of the octopus they were sending to countries like Spain, Italy, and Greece was not high enough to meet Japanese standards. If an octopus was missing an arm, had a lesion, smelled a bit of the gasoline where it had been tossed into the bottom of the boat, or had a bruise from rough handling, it would still find a buyer in Southern Europe. When an octopus was hauled aboard the boat of an artisanal Senegalese fisherman, it was likely to be killed immediately and tossed in the bow before the entire day's catch was brought to port. For this reason, Japanese importers preferred to do business with Mauritania and Morocco, where local fishers had learned and adapted to Japanese market requirements. But it was becoming evident that these waters were being overfished and that new sources of octopus would need to be found. In 2003, the Japanese international development consulting agency Overseas Agri-Fisheries Consultants Company sent a fisheries expert, Naohiko Watanuki, to Senegal, Mauritania's neighbor to the south.

"We Japanese eat a lot of fish and seafood. That is our national dish, and it must be in excellent condition," Naohiko Watanuki told me on a November morning in 2024 when I met up with him in a Tokyo café. "We are very strict, and Japanese seafood companies at the time were not interested in buying Senegalese octopus."

He was built a bit like a bear, if a bear wore glasses and spoke excellent and enthusiastic English. He had spent almost twenty years in Senegal, creating and promoting an octopus fishery before retiring in 2022. "I was very tired. All those years of traveling from Japan to Africa with no direct flight—Japan to Paris to Dakar, or Japan to Dubai to Dakar. Twenty-hour flights. I had taken enough long flights, and I was leaving my wife and children at home three hundred days a year. It was time to retire."

Like many Japanese men, he had a job that took him away from home for long periods. This is so common in Japan, where businesses are prone to transfer employees between various locations, that it has a name: *tanshin funin*, meaning "going to a new post singly." Now that he was retired, Naohiko had a plentiful supply of energy, and he liked to keep busy at home. When we met, he was in the process of building a guesthouse beside his home in Chiba, just outside Tokyo, and organizing a culinary school. His plan was that students from all over the world could stay at the guesthouse while they were taking his cooking classes and learning the secrets of *washoku*—"Japanese cuisine." "My life has gone from the sea to a guesthouse," he laughed, delighted.

On his initial 2003 visit to Senegal, he saw an abundance of octopus being caught, but the way the catch was handled, and the poor quality of much of what the Senegalese were already selling to Europe, meant he had a lot of work ahead of him. Because the Europeans would buy octopuses that were considered below Japanese quality standards, Senegalese fishermen were careless about how they handled octopuses. Naohiko Watanuki understood early on that he would have to spend a great deal of time teaching the Senegalese fishers. Yet he also saw that there was enough octopus to make it worthwhile for both the people living in small fishing villages and Japan's seafood companies. The Senegalese octopuses were small, and the Japanese liked smaller octopuses to use in takoyaki.

"I had two choices of where to start when I first got to Senegal: fisheries management or quality management," he recalled, pausing to search for just the English words he wanted. "I decided it was necessary to forestall overfishing, because the price for octopus was rising, so I started with what we call co-management. After that came quality management, quality improvement, step by step. It was a very interesting experience. What was important was that I took a participatory approach with the men who were fishing and the women in the fishing villages."

When he arrived in Senegal, sales to Japan were nil. "What I thought was that if Senegal could diversify its exportation to countries outside Europe, to Japan and Asia, it would be a good thing for both countries. Because octopuses in Senegal were caught by handline with an artificial lure—not nets or pots but handlines, caught one by one—it could be very selective and of potentially very high quality."

Now, in a good year, over two million pounds of Senegalese octopus are sold to Japan. "I worked very closely with the local people, something the European buyers did not do. I went to the fishing villages, and I went out fishing with fishermen. We ate together, drank together, and I slept at fishing villages. Most Western people do not go to the field. They just stay in the office in Dakar, Senegal's capital. Our working styles are different, and of course the Senegalese people prefer to work with Japanese. Our language skills may not be good, but what is important is to go the villages, to discuss with the people there how to solve their problems. We used their knowledge because they know very well about the octopus. Our approach is not top-down, but bottom-up. It all took me about fourteen years, but it worked.

"The seafood companies importing octopus to Japan were reluctant to buy Senegalese octopus because they had gained a reputation for poor quality. I worked on showing them how to have a better-quality octopus, and I went out fishing with them to show them. Before, when they caught an octopus, they would just throw it on the floor of the boat until they went back to port. Then, of course, it would smell of gasoline and whatever else. I showed them how to keep it in a net. When they brought the live octopuses to port, they would remove the guts and wash them with water. I showed them how to use ice. It's not good for the octopus to wash it with water; it needs to be gutted and kept on ice. Little by little we began producing high-quality octopus."

Naohiro Watanuki found funds to set up octopus pot manufacturing in Senegalese villages. He convinced the local women to make pots that could be sunk to the bottom in the fall, during spawning season, to give the female octopuses safe, sheltered places where they could successfully guard their eggs until they hatched. "It took about a hundred meetings—and all in French." He sighed and laughed. "Eventually some ten thousand pots were made, and it became a new income generator for the village women."

In addition to convincing the Senegalese fishers to change their habits, he worked hard at reaching the representatives of Japanese companies and changing their opinions of Senegalese octopuses. "After we improved the handling and the quality, we hosted an octopus party in Senegal at the Japanese embassy and served Senegalese octopus. It had a very good taste.

"It takes two months to send frozen octopus by ship from Senegal to Japan, but if you handle it right, there's no problem. We did a taste test in Japan with three octopuses from Senegal, Mauritania, and Akashi and invited the seafood company representatives and owners of takoyaki shops to try it for themselves. 'Okay,' they said, 'we can use it.' The octopus from Senegal was cheaper, but the quality was the same. In eastern Japan, they still don't eat so much Senegalese octopus, but in western Japan, in Osaka, they eat a lot of it."

Even the huge quantity of *O. vulgaris* arriving in Asia from Senegal and North Africa to augment the local catch was not enough to satisfy demand. When Japanese seafood companies began looking for other suppliers, Jesús Gutiérrez Aguilar was waiting for them. *O. maya* from Mexico had never reached Japan, but he was convinced that his family seafood business, Promarmex, in Progreso, could provide a product with a taste and texture that would meet with Japanese approval. He began to contact various Japanese seafood companies to find one willing to try marketing octopuses from the Yucatán.

To visit him when I was staying in Sisal, I had to make a fifty-mile road trip to Progreso, even though only about twenty miles of coastline separated the two towns, because Hurricane Gilbert had destroyed the road between them in 1988. It took almost four hours in two vans: The first van, going inland from Sisal to Mérida, stopped in the town square at Hunucmá for a while before going on to Mérida, where I waited for the second van—a decrepit vehicle with cracks spiderwebbed across the windshield. The second van eventually made its slow, crowded way to Progreso, and once I reached the Progreso bus terminal, I took a taxi to the Promarmex factory up the coast.

The driver was a square-shouldered, middle-aged woman. "I'm going to drive a little fast," she told me. "But don't worry. I know what I'm doing. The thing is, after I drop you off I have to go home, because my sister's there; she's sick. She's had dengue for a week, and she's not getting much better.

She'll be okay after a while. Her fever's gone down a little, but it's taking time. You get that dengue, you're probably not going to die, but you may want to. It's no fun."

She dropped me at the factory, and Jesús Gutiérrez met me and brought me upstairs to his office. Promarmex was founded by his grandfather, and his father followed as head of the company. They began by selling strictly to a national market, buying the local sardine-like fish called charales from the fishermen and laying them out to dry in front of the factory, then packaging them, dried and salted, for sale on the national market. Dried charales are eaten as a snack food in numerous parts of Mexico. The company built huge ovens and turned some of the dried fish into fish meal. Eventually, under the administration of Jesús Gutiérrez's father, the company expanded into frozen fish and seafood, but its market was limited to Mexico and some sales of fresh fish to the United States. Jesús began working with his father in the early 1980s at the age of twenty.

Until the mid-1990s, people in the octopus-consuming countries of Asia had never eaten *O. maya*, and Japanese seafood buyers had assumed that it would not satisfy their customers, who were used to the taste of *O. vulgaris*. Jesús Gutiérrez was the first of the Progreso processors to start changing that opinion and building a commercial relationship with Japan. "In those days octopus wasn't worth anything, really," he told me. "Most Japanese had never even heard of *Octopus maya*, and Europeans were not interested in it, believing the common octopus to be the best-tasting in the world. But as demand rose, and as supplies shrunk in the mid-nineties, *O. maya* began to attract interest. We were one of the first to export to Europe. We received a permit from the European community in 1996, and after that we began selling to Japan."

Jesús Gutiérrez initiated contacts with Japanese companies and subsequently made the first sales of frozen *O. maya* to Asia. The Japanese company that reached the first agreement with him sent an employee to oversee the operation. The animals had to be kept in ample cages, allowed to relax, killed quickly, without suffering, and then frozen on the same day they were caught. Only if those conditions were met would the company buy the octopuses.

"You need a permit to export to Europe, but you don't need one to sell to Japan," he told me. "Even so, they insist on a higher quality than any other market. Their standards are much more difficult to meet. They are by far the most difficult to satisfy. After that come the Europeans who are less demanding, and it's easier still in the United States."

The business his grandfather started is now an international player with a fleet of thirty trawler-sized boats and numerous artisanal fishers, employing some five hundred people during octopus season. The frozen *O. maya* is transported in container ships and delivered directly to a plant in Europe or Japan. Much of their export to Japan consists of precooked frozen octopus, which will be used in takoyaki.

Jesús Gutiérrez was surprised and gratified that by 2025, Mexico's neighbor to the north had become an important consumer of his octopuses. "We've begun to sell as much octopus to the United States as to Mexico. Not too many years ago, it was difficult to find octopus in an American seafood restaurant. Less than one in every ten. Now it's more like ten for ten. We sell to many Italian restaurants, and to restaurants all over Los Angeles."

Promarmex may have been the first of Progreso's seafood companies to contact Japanese and European buyers, but its success immediately spawned competition. In 2024 Progreso was home to some twenty octopus-processing plants, all working full bore during the six months of octopus season to provide product for consumers around the world. At Promarmex, during the busiest months, the plant ran around the clock, with three eight-hour shifts. Even when octopus season was over, the company aimed to have enough frozen *O. maya* on hand to send their customers until the next fishing season came around.

The rapid spread of sushi restaurants in both Europe and the States has increased pressure on the world's octopus stocks. Google Data Trends, according to an article in *USA Today*, found that sushi was twice as popular in 2023 as it was five years before and more than three times as popular as it was in 2013. "Trendy sushi restaurants now exist in every metropolitan city in the country, and nearly 5 million Americans eat the dish at least once a month," noted the newspaper.[17]

The steadily increasing demand for octopus around the world has generated a growing problem with false labeling. Not many of the foodies in New York City and Los Angeles who are devoted to eating octopus would be able to tell whether they are eating *O. vulgaris* or the cheaper *O. maya*. Purists will insist on being served the common octopus, but they are unlikely to know if they are served a four-eyed octopus from the Gulf of Mexico rather than an animal from the Atlantic Ocean.

Vince Cutrone has been a wholesaler of octopuses in New York City from his shop, the Octopus Garden, since the mid-nineties, and his customers include some of New York's finest restaurants. A few years ago he had to move from Brooklyn to a larger shop in Staten Island in order to accommodate the growth in business. He insists on only dealing in *O. vulgaris*, although he concedes that there's not much difference in the taste from *O. maya*. Nevertheless, his customers expect *O. vulgaris*, so that's what he sells them. "We're not having supply problems. Our suppliers complain that the yields are not there, but if you're willing to pay for quality you won't have any problem finding it."

Cutrone is a short, trim man, accustomed to rising at 5 a.m. to prepare the day's deliveries into Manhattan. The octopus he loads onto trucks at dawn could be on the plates of some of New York City's most exclusive restaurants that evening. But he also has a retail business and a steady roster of retail clients, mostly Italians, who want octopus to grill at home. Vince Cutrone arrived in the United States in 1974 at the age of fourteen from Bari on the Adriatic coast of Italy, and both he and his customers like to eat it the way it is served there. Every July Fourth he sets up a large grill in a park and grills octopus until his supply runs out.

"They line up for it," he told me. "I make my grill very hot. The octopus has the guts and the beak taken out, and I open it up and put it down on the grill pretty flat. When it changes color and I see it's done, I put it in a good olive and lemon juice, and it draws them up. It's delicious. You can even make sandwiches with it."

Even for his outdoor celebration, he would never think of using *O. maya*. "I've tried it," he told me, "and I'm not really sure there's any difference in taste. Maybe *vulgaris* is a little firmer. I don't know, but my customers want *vulgaris*, and the customer is always right."

Of course, not everyone has Vince Cutrone's scruples. Ian Gleadall was the lead author of a 2023 paper about traceability of cephalopods from origin to table, which concludes that "meaningful traceability of cephalopods [is] a prerequisite for reliable, effective and meaningful monitoring and regulation of sustainable production." An estimated 80 percent of the octopuses coming from Mauritania have been caught by artisanal fishers, and the report finds that the numbers they provide for stock assessment, if they provide them at all, are often unreliable. A lot of the octopus imported into the US is from Mauritania, but researchers state that "the route of this harvest through to arrival in the U.S., and the monetary value obtained by the local Mauritanian fishers, are currently obscure."[18]

The study noted that one 2023 survey at an Italian port found that 44 percent of the cephalopod products, including those labeled *O. vulgaris*, had been misidentified, and a 2020 survey of three supermarkets in Spain revealed that a whopping 66 percent of the cephalopod products were improperly labeled. This happens for a number of reasons—some of them honest mistakes between two similar species, but other times it is intentional in order to obscure the origin of animals fished illegally or to substitute a cheaper product for a more expensive one. In 2022, Ian Gleadall's Spanish coauthors found a number of supermarkets in southeastern Spain selling jars and cans of things like "octopus pieces in garlic sauce," which were actually squid.

Stock assessment is important. To have an accurate idea of the sustainability of a given stock of octopuses it is necessary to know how many of them are being landed. The study's authors conclude that current methods are inadequate to understand where octopuses on the market came from: "To accurately identify cephalopod seafood products requires a reliable, affordable, and widely accessible system capable of tracking seafood from its point of harvest through its distribution," the study stressed.

Several concerns about the lack of traceability for octopuses along the global marketing chain would be addressed by farming octopuses, which would be more easily traceable. Of course, a new issue might arise if farmed octopuses were put on the market and labeled "wild" in order to charge more for them, as is known to happen already with salmon.

Regardless of where they originated, when an octopus of decent quality is offered for sale, there is no shortage of willing buyers. In addition to the Japanese eager for top-quality octopuses, other countries are in line for the animals that don't meet Japanese standards. Between 2005 and 2016, Spanish imports almost doubled, rising from 31,809 tons to 55,148, with the catch by Spanish boats able to cover only a quarter of the total consumed, according to the Spanish Chamber of Commerce.[19]

The world depends on places like Mexico, Mauritania, Senegal, and Morocco for its supplies of octopus, but the seasons and quotas for their capture in these places are written in butter, not in stone. A season can suddenly be closed or a quota imposed depending on the condition of the stocks. In Morocco, for instance, 2022 was generally viewed as a stellar year, 2023 was just so-so, and in 2024 the opening of the winter season was postponed by a couple of weeks because of low numbers of octopuses. Li Yemin complained to me that in Mauritania she had to spend a lot of time keeping abreast of seasonal variations in the allowable catch, and in Mexico, a red tide sometimes imposed a sudden suspension or even closure of a season. All of this, in addition to the important variations in octopus populations from year to year, meant that the market was unstable. It was difficult to ascertain the real condition of the world's stocks. In a few decades octopus has gone from being a locally caught, inexpensive dish for people living in coastal areas to being the driving force behind a vast and lucrative global market. How will farming them fit into that trajectory?

Will we ever be able to farm them? And even if we could, should we?

Chapter 5

COULD WE?

I WAS driving back to my home in Barcelona, pondering these questions after a few days hiking in the Pyrenees, when I passed a string of trucks hauling trailers full of live pigs to the slaughterhouse. As I pulled alongside each live-haul trailer, I could see the full-grown animals with their wet, pink snouts pushed through the rusty iron bars. Western Catalonia is pork country, flat and full of hog farms. The air was redolent with the smell of porcine waste, as it always is on that stretch of highway. For many Catalonians, it was not the pungent, unpleasant smell of pig feces: It was the smell of money.

And it was the smell of supper. So deep is the love of pork in Catalonia, and all over Spain, that as long ago as 1492, when church inquisitors wanted to know if a person had truly left behind the Jewish faith and converted to Catholicism or whether they were merely pretending to have converted, the converso was presented with a dish of pork to see if he or she would eat it. A neighbor's word that someone would not eat pork was enough to have him or her detained, brought before the Inquisition, and presented with a plate of ham. This would determine if the conversion had just been a ruse to save them from punishment, which could have included execution, or if their new faith was real. Sometimes, what we will and won't eat is a life-or-death choice, but most of the time it's less important, although it may be deeply rooted in our personal and cultural identities.

As it happens, those long stretches of highway enveloped in the scent of pig excrement were the result of factory farming rather than something intrinsic to the nature of swine. All that accumulated waste from many pigs in one place, gathered together in a pool or a holding tank, creates the pervasive smell. A pig in a properly maintained pen in the backyard is not likely to have an offensive odor. In fact, pigs have been shown to have considerable similarity to *Homo sapiens*, perhaps more than octopuses. Pigs have large brains, as we do, and a genome that is not so far removed from ours.[1]

Pigs, like octopuses, have a surprising capacity for learning, according to a 2021 article: "Over the past few decades, research has demonstrated pigs' capacity to comprehend symbolic language, plan for the future and discern the intentions of others. Studies have found them to rival chimpanzees in their ability to learn and play joystick-operated video games, despite the fact that their feet and snouts are inevitably less adept at handling the mechanisms."[2] Similarly, a 2007 study showed that after being exposed to an object for two days, pigs still remembered it after five. Another study showed that when given a choice between two food sources, one with more than the other, the pigs remembered the one with more food and consistently chose it.[3]

Pigs are enough like us that kidneys from genetically engineered minipigs have been implanted in primates using CRISPR techniques, and the monkeys have survived more than a year. Next up: humans. In 2024, in Massachusetts, the first such transplant was done; two weeks later the patient was sent home to enjoy the rest of his life with his new kidney. Unfortunately, that proved an extremely short time—within two months the patient had died. Nevertheless, many medical researchers believe that genetically engineered pig parts will one day soon serve as organs for humans. Numerous experiments transplanting pig organs to other animals have encouraged researchers to posit that such operations, called xenotransplants, will become a reality in the not-too-distant future and that this kind of cross-species transplant will go a long way toward solving the shortages of donated organs around the globe.

Many of the people who refuse to try octopus meat because they view the animal in a sympathetic light benefit from research done on and with pigs. Many also regularly eat pork even if they know the pigs come from one of those factory farms that offend any nose passing by on the highway—not to mention any ethical sense about how animals should be farmed. Diners need only look at one of numerous clandestinely filmed videos online showing the interior of industrial pig farms to see the nasty conditions under which the animals are born, raised, and slaughtered. People still eat a lot of pork.

If farmed octopus eventually reaches the market, the first challenge will be convincing people to try it. Once meat eaters taste an octopus that has been correctly prepared for the first time, they will usually be happy to eat it again, but they have to be willing to try that first bite. Could that possibly happen on a wide scale?

Taste is one of the most difficult human senses to account for. What people are willing to eat is by no means universal, nor always readily explicable. It was ever thus. In Greek and Roman cultures, while octopus and fish were widely appreciated, meat eating was generally looked down on by people whose diets were made up primarily of fruits, vegetables, and seafood. The German and Celtic societies had the opposite view, considering meat to be indispensable and disdaining plant-based meals, considering them to be the default food choice of poor people who could not hunt well for themselves nor afford to buy meat.[4]

Studies have shown that some people are more inclined than others to try new foods. One study among 265 Sicilian consumers found that there were two types of reasons for refusing to try something new. One was neophobia, which was simply a dislike for anything new, outside of the consumer's cultural norms. The second motive was food disgust, in which a new food elicited a negative reaction, a rejection that could be based on cultural values but which might also have psychological roots or even be tied in to our evolutionary means of defense. An unknown food might actually be dangerous to us, the reasoning goes: It might be poisonous, carry pathogens, induce illness, or cause nausea.[5]

Foods like eel or octopus are often viewed with disgust by people who have never tried them, simply because they look "weird." When the average North American supermarket shopper sees an octopus splayed out on ice, it is initially a stretch to imagine how it might be cooked and served. Even when new foods are not considered dangerous, the way they look is often enough to elicit a negative response. The very word "disgust" comes from a root meaning of "bad taste."

I myself am not immune to disgust when confronted with a new food, a fact I discovered in Ecuador, a country whose citizens, like people all over the globe, believe the culinary specialties of their homeland are the tastiest in the world. This belief is a little hard to swallow when one is confronted with the grub larvae avidly consumed in Misahuallí, a town in the Ecuadoran Amazon jungle. It's a typical, small Ecuadoran jungle river town, a population of nearly five thousand, three-quarters of whom are Indigenous ethnic Quechuas. When I was there, Misahuallí had two blocks of commercial establishments around a main plaza, a small square of green in the center of town with a few benches and palm trees. On the other side of the asphalted street that runs along the plaza's perimeter was a raised sidewalk, backed by stores and cafés, a handful of clothes shops, food stores, and storefronts that try to entice the occasional tourist by advertising boat trips on the Napo River. Among the business establishments were three typical cafés with the same $2.50 menus—soup, a meat and rice course, drink, and dessert. The same kind of cafés and menus are found all over Ecuador—except for the skewers of fat, shiny, pus-yellow grubs called *chantoro* on grills outside the door of each Misahuallí café, enough skewers grilling to signify that it was a popular local dish.

I stood gazing down at them, thinking about whether I wanted to put one of those grubs in my mouth, and marveling at the sensation of disgust welling up inside me. The café's manager was quickly at my side to explain that they were beetle larvae, harvested from a local palm tree, and he promised me that they were delicious. He urged me to sample one, recommending it for both its flavor and health benefits. I gave it some consideration but couldn't bring myself to put one of those grubs in my mouth. Nevertheless, I noted that the locals dining at the tables inside were not having any problem doing so and were evidently relishing it.

Another new food I found all over Ecuador and did try was guinea pig, which is consumed with gusto across the country. Somehow, guinea pig did not arouse more than a modest initial disgust, which quickly turned to a shrug of the shoulders and ordering one to try it. Like the chantoro larvae, it is typically skewered and cooked on a grill. The sight of a skinned guinea pig, called a *cuy*, splayed on a skewer and roasted over hot coals is not the most appetizing, particularly for someone whose second-grade class had a guinea pig in a cage in a corner of the classroom, much loved by us students. Perhaps it was the vague resemblance of the skewered guinea pigs to chicken that made it seem more palatable to me than the grubs, but I did give it a try—and it wasn't bad, with a flavor leaning toward Cornish hen.

Research done by Paul Rozin and colleagues concluded the only thing that universally disgusts human beings when considered as food is excrement.

And that's only when eating it directly. Pigs and dogs are both considered food in some places, and they both have no problem eating feces. Other than the universal human reaction to chowing down on caca, Rozin and colleagues posit three different motives that elicit disgust when people are considering a new food. The first is when something has an undesirable sensory property, like taste or smell. This is the only one of the three that is present at birth and is most associated with bitter taste, which consistently elicits a reaction of disgust and rejection from even the youngest of babies. Bitterness is a taste we have to learn to love in order to overcome our initial disgust. Sweet, on the other hand, is a taste infants like from their earliest meals.

Another reason we may find a food disgusting is because it is dangerous and has unpleasant physical consequences, like diarrhea or nausea, when it is eaten. After the first such experience, people are unlikely to try the offending food for a second time. We come into the world ready to try putting almost anything into our mouths. One of the first lessons a parent must impart is what is safe to swallow—infants and young children want to taste just about any new object they come across. Every parent has had to tell a young child not to put something in his or her mouth, probably many times. Paul Rozin writes, "In the first years of life, children learn that some substances are not good for them. They either directly experience discomfort upon eating these substances or they learn from others to expect such discomfort."[6]

Children in developed countries may be the pickiest eaters on the planet because they have a wide range of reliably available choices about what to eat when they are hungry. The transition from children willing to put anything in their mouths to insisting on peculiar and, apparently, unnatural food habits can happen in a remarkably short span of time. By the time she was a toddler, my granddaughter refused to eat anything that did not wholly occupy a single space on a plate. Everything in its place and no overlap. Mixed salad? Forget it. Rice and vegetables? The rice had to be by itself, as did each separate vegetable, and if something crossed those borders she was genuinely disgusted and wouldn't eat it. She lost this obsession by the time she was eight or nine.

The pickiest eater I have ever known was a young friend of mine in his early teens who I'll call Will, from a well-to-do Swiss-Scottish family. Will would only eat "white" food like white bread, spaghetti, mashed potatoes, white rice, and not much else. This went on for years, and he stood his ground. He could be hungry, but if you tried to serve him beef stew, or a peanut butter sandwich, or anything else that wasn't white, he would shake his head in denial. It didn't seem to do him any harm; he continued growing into a tall, strong guy, but he held fast to his eating principles right through adolescence. When I phoned him decades later to ask what had motivated him to stick to such a weird diet, he mentioned that at thirty-one years old, he had outgrown those early eccentricities. Then he mused, "I'm not really sure. That [eating only white foods] was always the norm for me. I would only drink milk and eat loads of rice or pasta or white bread. I would eat all the time, but I was quite picky. It wasn't that other things disgusted me, but even today, when I'll eat almost anything, if I find something I like I'll just repeat that. I never felt an urge to switch, and I preferred having that to anything else. There was no medical reason; I didn't have any allergies or anything like that."

Will has always been a self-aware person, and that played a role in how he grew out of only eating white foods. "When I got to university I began to diversify, partly because it seemed to me to be kind of a snobby, privileged thing to say, 'Well, I'm only going to eat white things,' and I thought it was not a very good trait, so I began to consciously change. Then in 2015, I moved to China and lived there for a year, and then I ate everything from snakes to crickets. On the one hand because it was all delicious, and on the other hand because I thought being a picky eater was stupid of me, like not being willing to try different things."

What we eat is clearly bound up with the image we have of ourselves. "In adolescence and in adulthood food choice can be a reflection of self-image; conformity or individualism can be displayed through eating behavior. What you buy and what you eat tells others that you are . . . thrifty or extravagant, modern or old-fashioned. In this way food choice becomes a manifestation of personality," wrote the Canadian health promotion specialist Paul Fieldhouse.[7]

All these things related to taste—self-image, neophobia, belief systems, access to resources, and simple pickiness—could create obstacles to potential consumer acceptance of octopus meat. However, a good deal of research has been conducted about how to convince a person to change eating habits, and the decision to try a new food is frequently based on numerous factors. A 1979 study elaborated a list of twenty reasons why people eat what they do; only one of the twenty was to satisfy hunger and provide nutrition for the body. The other nineteen were sociocultural reasons including demonstrating love and caring, rewarding or punishing behavior, expressing moral sentiments, signifying wealth, or demonstrating belonging to a certain group.[8]

Any effort to introduce new, untried food to people has to take such socio-cultural motivations into account.

A new food may acquire acceptance from publicity about its benefits: A previously untried food may come to be seen as promoting health or weight loss or even having aphrodisiac qualities. Another source is the influence of people we trust. Not surprisingly, a 1986 study showed that college students were less likely to accept a new food offered by a stranger in the subway than by a friend in the friend's home.[9]

Sometimes, a food's social status may be enough to entice someone to try it. If the new food is something that previously only showed up on the plates of the rich and was out of reach for the poor, it may be attractive if it does finally become more widely available and affordable. Potatoes were originally reserved for the tables of Europe's elite, and the same trajectory was followed by coffee. Initially these foods were scarce and therefore costly, but as they became more widely available their prices came down. People who had watched the rich enjoy these things were ready to try them.

All of these factors will come into play if a decision is made at some point to try and introduce octopuses into the national diet. It would not be the first time that efforts were made to control what someone is willing to eat. Food manufacturers, marketers, governments, and supermarkets have all studied ways to change eating habits. In one 1975 study, two out of three people said that they were willing to try new food they had seen advertised on television, and one out of two people were willing to try a new food that looked attractive on a supermarket shelf.[10] Other, less obvious factors may also be at play when people decide to try something new. A 2021 study in the journal *Personality and Individual Differences* concluded that "the willingness to try new food can influence sexual desirability and is seen by others as a signal of sexual unrestrictedness."[11]

In 1941, almost a full year before Pearl Harbor and the United States' entry into the world war, Washington officials began expressing concern about national dietary habits. Not, as one might guess, in order to prepare for food rationing, but rather because there was a concern that bad nutrition had a deleterious effect on the physical fitness of the military and the citizens at large and that one aspect of national defense should pay attention to the nation's physical fitness.

The National Research Council founded the Committee on Food Habits and staffed it with a group of prestigious researchers to study ways in which the habits could be changed and how people might be convinced to adopt a more nutritious diet. The noted anthropologist Margaret Mead was named executive director. In Carl E. Guthe's essay about the committee's creation, he observed that it expressed particular concern about the lack of good nutrition among "the low income non-farm groups, largely concentrated in the eastern Industrial Belt; and the low income groups of the Southeast, including especially the farm and negro population thereof."[12]

In Margaret Mead's 1943 essay on the committee's findings—still relevant today—she concluded:

> The long-time task is to alter American food habits so that they are based upon tradition which embodies science and to do so in such a way that food habits at any period are sufficiently flexible to yield readily to new scientific findings. In order to accomplish this goal, the food habits of the future will have to be sanctioned not by authoritarian statements which breed rigid conformity rather than intelligent flexibility, but by a sense of responsibility on the part of those who plan meals for others to eat. At the same time it will be necessary to invent channels through which new findings can be readily translated into the meal planning of the woman on the farm, in the village, and in the city. To devise such a system of education, communication, and change which will link the daily habits of the people to the insight of the laboratory, and at the same time contribute to the development of a culture which produces individuals who are generally better adjusted as well as specifically better fed, is a task which requires a recognition of the total cultural equilibrium.

Even those changes would fall short, she warned, if the food distribution system, and patterns of consumption, did not change:

> Further, the application of science to improve eating habits may become empty and meaningless if it is not paralleled by efforts to apply science so as to increase the supply and adequate distribution of food. Efforts to better the nutrition of the world simply by altered production and distribution will also fall short of their goals unless corresponding and congruent changes are made in the patterns of consumption.[13]

This was written before fast-food chains existed and when obesity was not the widespread problem it is today. And it was a time when a much larger percentage of Americans were used to contributing some part of their own diets with vegetable gardens or domestic livestock. Eighty years later, in a nation dealing with a serious problem of obesity, the question of how to change the North American diet is more urgent than ever.

Among its research projects, Margaret Mead's committee carried out a study in "a Midwestern town with a population of about sixty thousand" in 1942 to determine how effective a group discussion and group decision process would be, compared to a lecture, in order to convince housewives—who were almost always the family meal planners—to cook and serve a supper of unfamiliar organs like beef hearts, or kidneys, or brains. Participants who opted to do so were provided the organs free of charge and given recipes for preparing them.

Initially, to the 107 participating homemakers, these organs were about as appetizing as drinking water out of a toilet. However, many more in the discussion-and-decision group eventually served one of those organs than those who were urged to do so in a lecture. As many as 50 percent of the homemakers in the discussion-and-decision group eventually served their families something entirely new, while only 8 percent of those who only heard a lecture did so. The single most convincing theme in both groups was describing the meats as beneficial to health. This was true regardless of income or race. When people believed that health benefits would accrue from eating a previously untried food, they were most often willing to do so. The study's author concluded, "These experiments suggest that it may be possible to bring about in a relatively short time definite changes in food habits, even in food items which would be expected to show great resistance to change."[14]

Group discussions in the digital age are easier to convene than ever and communal decisions easier to take. These days we can, and do, organize group discussions and decisions without participants even having to leave their homes. It is easy enough to bring consumers together online. The marketing department of the first company to put farm-raised octopus on sale will want to make its health benefits—no fat, lots of antioxidants, high in omega-3 and protein—stand out. Social media influencers will need to be recruited.

Many of the foods that form a part of what we think of as a basic diet were not always accepted as such. New foods generally are slow to be adopted, first being eaten by a few people. Or at times when conditions like food scarcity or famine encourage people to look outside the normal sources of nourishment. The potato, for instance, came to Europe with sixteenth-century Spanish explorers, brought back on ships returning from South America. It took three hundred years in some places before it became the staple that it is today. The first potatoes were referred to as "Eve's apple" or "Earth's testicle," hardly nicknames to attract people to try them. In France, it was not widely accepted until the nineteenth century, initially regarded as insipid in flavor and thought to cause flatulence.[15] Tomatoes were considered aphrodisiacal or poisonous for centuries in Europe following their introduction by Spanish explorers.

For centuries, the rich and the poor had different diets. Some things, like garlic, onions, beer, beef, beans, and soup, were seen as best fitted to a poor person's constitution because they needed "heavy foods" to fuel the hard work they were expected to do. A lighter fare of young animals, pigeons, airy desserts, and fish were the sorts of food that might be served to the rich or at court dinners.[16] Little has changed. The rich often eat things in the world of haute cuisine that the poor will never try—dishes created in laboratories where chefs spend their time elaborating things to eat that don't resemble food as we know it or, if it appears recognizable, turns out to be something entirely different. The common denominator for these unusual dishes is a price tag far out of reach for an average working person.

I was once sent to the three-star Michelin restaurant Arzak, paid for by a glossy magazine that hired me to go and write about the plates and pleasures of this world-famous eatery just outside of San Sebastian, in the Basque region of Spain. I had a $250 tasting menu reserved for lunch at the restaurant's two o'clock serving. I arrived early in my rental car, right around one o'clock, an hour too early for a midday meal in Spain. The dining room was empty, although the front door was open. I carried on through to the kitchen where I saw that the Arzak family and the kitchen staff were having lunch. Seated around a big, round wooden table at one end of the kitchen were the third and fourth generations of a family-owned restaurant open since 1897—Juan Mari Arzak, the chef who gained its first three-star Michelin rating, and his daughter Elena, who was carrying on the tradition, along with a half-dozen employees.

Aha, I thought as I approached the table to introduce myself, *let's see what delicacy the Arzak family has prepared for their own lunch.* In the middle of the table was a huge pan of macaroni and cheese, which everyone was spooning onto their plates and forking into their mouths with great gusto. At 2 p.m. I sat down to the tasting menu, and while the small dishes were clever— exquisitely presented and tasty—I couldn't help yearning for a plate of that macaroni.

Aquaculture can level the gourmet playing field a bit by providing afford-able access to foods previously restricted to the well-to-do. This is what hap-pened in the cases of both salmon and shrimp. From 1982 to 2007, salmon farming numbers around the globe grew ten times larger, and by 2022 salmon aquaculture was worth $20 billion a year.[17] When only wild-caught fresh salmon was on the market, it was an expensive fish, growing ever more so as its numbers declined, but thanks to farming it became available at more reasonable prices. Salmon is currently one of the most consumed fish in the world.

Salmon farming brought the fish to lots more dinner tables—good news for consumers, but it is fraught with some negative environmental reper-cussions. Most aquacultured salmon is raised in open-net ocean farms, where the fish may take a couple of years to grow to market size in densely packed net cages below the surface, which are anchored to the bottom. The fish are vulnerable to diseases and parasites—particularly to sea lice, which feed on and weaken the live salmon. Sea lice account for a large share of the mortality among farmed salmon, along with various diseases. All too often, these are treated with chemicals and antibiotics that can reach con-sumers through the salmon's meat; the residue that drops to the ocean floor eventually makes its way into other seafood and fish that people eat. The farm-raised salmon on our plates also has a significantly higher ratio of fat to meat than its wild counterpart because the fish has less opportunity for exercise during its lifetime, restricted to the boundaries of the net cages. Some salmon may escape the open-net pens, and if they are infested with sea lice or carrying diseases, these may spread to the wild populations.

Another major problem with aquaculture is that the farms are likely to use smaller fish for feed, which is where a salmon gets the protein neces-sary for its growth. An estimated 20 percent of the world's wild fish capture currently goes to feeding farmed fish. Fish that might otherwise wind up as a nutritional meal on a family's table are being vacuumed up in industrial quantities to be transformed into aquaculture feed.

Shrimp, rich in protein and consumed in various ways around the world, is another food that used to be associated with high-end seafood restau-rants. Sixty years ago, most Americans had never eaten shrimp, but in 2020 the average American consumed five pounds of it, up 25 percent from 2015.[18] By 2016 55 percent of global shrimp consumption was farmed in an aqua-culture worth billions of dollars, mostly located in underdeveloped coun-tries.[19] Nueva Pescanova, for example, is deeply invested in shrimp farming, annually raising some sixty thousand metric tons of vannamei shrimp—also known as white-legged shrimp—in Ecuador, Nicaragua, and Guatemala.

Among shrimp aquaculture's most common environmental problems are pollution by wastewater from the farms and destruction of habitats. Shrimp farms are often constructed in mangrove swamps, salt ponds, or wetlands, sometimes leading to their pollution or destruction. In addition, chemicals and antibiotics are often used in shrimp ponds, and these can have negative effects on the health of animals or humans who eat the shrimp. Despite the chemicals and antibiotics, viral and bacterial diseases threaten shrimp in many places. Diseased shrimp may be eaten by waterfowl, which then spread the disease to other shrimp farms.

The most frequently farmed fish in the world is tilapia, originally native to Africa, now farmed in more than ninety countries. It is easy to raise, hardy, and tolerant of relatively harsh environmental conditions. Tilapia also shares the problematic environmental aspects of shrimp and salmon farming, augmented by the fact that regulatory oversight in many of the countries where it is farmed is often ineffective or nonexistent. Tilapia has only moderate nutritional benefits and not much taste, but it has put cash in the pockets of many small farmers, along with placing fish on the supper tables of many families who could not previously afford it.

While the cuisine of well-off people sometimes works its way down to the tables of the lower classes, the reverse also happens, in which foods once found only on the plates of the poor or working class move up in status. The food then becomes expensive and out of reach of the people who used to eat it. One example is "elvers," young eels. These juvenile eels are about as thick as a human hair, semitransparent, and a couple of inches long, with two black dots of eyes at one end. Elvers leave the saltwater in which they were born to enter rivers, where they will grow into big adult eels and live fifteen or twenty years until heading back downriver to the ocean, returning to their birthplace in the Sargasso Sea, where they mate, reproduce, and die.

A hundred years ago elvers were so common in Spanish rivers that people slopped their hogs with them. But food for humans became scarce during the two decades that followed the end of the Spanish Civil War in 1939. People suffered real hunger, but elvers were still plentiful during the three-month season when they showed up to enter the rivers in great numbers: Walls of them rode the tide into freshwater on dark, moonless nights. They were slowly redirected from pig pens to dinner plates. In Agiñaga, Spain, in Basque country near the coast of the Cantabrian Sea, no one is certain who was the first person to stop throwing the elvers to the hogs and toss them into a frying pan instead, but there is general agreement that it was probably someone poor and hungry.

"It was a group of our grandfathers from Agiñaga that first began to eat elvers, and it was my father who began to sell them to restaurants here and began the first business with elvers," Santiago Otamendi told me. He was the sales director of El Angulero de Agiñaga, the business founded by his grandfather, one of five family-owned elver businesses in the Basque village. "My father used to go down south of Tarragona to the Ebro delta, and in one night there they could catch five hundred kilos of elvers. There wasn't anyone else doing it."[20]

Now elvers are scarce; they are on the European list of endangered animals and way too expensive to serve at home. They are offered rarely in restaurants, mostly high-end places in Basque country during the seasons when the elvers enter the rivers. When on a menu, they are priced at about a hundred dollars for a one-hundred-gram serving of *angulas a la Vizcaina*, elvers cooked in the Basque style, which means flash-fried in olive oil with nothing more than a little garlic and the mildly piquant *guindilla* pepper. Cooked like that, they are quite tasty, like a slightly grainy pasta with a tiny crunch from the backbone followed by a faint fishy aftertaste of the sea and garlic, along with the pepper's light bite. The Basques that go out with their fine-mesh nets to capture them at night in the rain and cold of the winter—elver season—cannot afford to eat what they catch.

Another poor people's food from the sea now sought by the rich is lobster. The early North American colonists were already familiar with lobsters when they landed in New England, and they found so many of them in North American coastal waters that they became the thing you would eat if unable to afford, or find, anything else. William Bradford, governor of the Plymouth Colony, complained in 1622 that food was so scarce he was forced to serve his guests only lobster.[21]

That has certainly changed. Among North Americans, a lobster dinner is generally a sign of middle-class membership. Poor people don't often eat it. Each lobster is worth so much that many lobster fishers—like their counterparts netting elvers in Basque county—won't even keep enough out of a catch to feed their families, preferring to sell all they have and put something less expensive on their own tables.

The whole ethos of lobster is something of a joke to anyone who has ever gone out on a lobster fishing boat. The crustacean is caught in traps off the New England coast, mostly by artisanal fishers, and when I went out in 2002, the bait most often used in the traps was skate. Each lobster boat carried a big, blue, plastic barrel packed with dead and rotting skates, the more putrid the better. We're not talking about slightly "off" here but rather disgustingly rotten. After a day on a lobster boat, it takes more than one shower to get the smell of the bait out of your hair. This preference for dead and rotten initiates a lobster's marketing trajectory, which will end by putting this vulture of the sea on a plate at an upscale restaurant where a diner will pay dearly to eat it.

Octopus, too, has followed the trajectory—like elvers and lobster—from being a basic food to a luxury. For most of human history, eating octopus was only done by those who lived near a coast or who bought cheap octopus from those who did. Octopus was a meal that could be obtained from the immediate environment, and the dish was not particularly valued above that. At the same time, octopus was no less appreciated for its flavor. It was, and is, a tasty meat. Perhaps, with the right advertising campaign, digital media influencers at work, and image makeovers, folks in Middle America could come to echo the Galician campesino who praised a dish of octopus with unbridled enthusiasm in Luis Villaverde's book *Mariscos de Galicia*: "It doesn't have bones, or splinters, or shells, or seeds . . . it's all meat! And what meat!"[22]

So noncoastal North Americans could eventually be convinced to buy octopus and incorporate it into the family diet. But should they?

Chapter 6

SHOULD WE?

THE SURPRISING thing that happens in a one-on-one with an octopus is that when you make eye contact, it makes eye contact in return—it looks back at you. That is, as you observe an octopus, it is observing you, and it's hard to escape concluding that it is also, in some way, considering you, *thinking* about you. And when it puts an arm out to lay a row of suckers on your own arm, it is not an aggressive act but a means of gathering information to add to that consideration.

My first time in the same space as a living octopus was on a Catalonian fishing boat in the Mediterranean, in 2007. The octopus at my feet, on the deck of Pere Pau Gras's boat, was a ruddy reddish-brown, a clear sign that it was feeling exceedingly threatened. It was making its way around the aluminum deck of the boat, moving in an odd kind of dignified, hunched slither, with one of its eight arms extended in front of it, feeling around this completely new world like a blind person with a cane, processing information as it explored, searching for a way back to the sea or a place to hide. Suction discs lined its rubbery arms. Its head was a bulb with protuberant eyes and a beak in the middle of the membrane that connected its arms. A gelatinous sac, which held its entrails, trailed along behind the head.

Octopuses have a justly deserved reputation among zookeepers and researchers as skillful escape artists, and many are the tales told of how they squeeze that bizarre body mass into, out of, and through what seem to be impossibly small spaces. None of that skill would help the angry octopus at my feet. Pere Gras was a short, stocky, thirty-nine-year-old Catalonian who had been fishing octopuses out of that patch of the Mediterranean for fifteen years, and his twenty-seven-foot boat was escape-proof. At his leisure, Gras would recover the octopus he had tossed on the deck for me to observe and add it to the contents of a long net bag suspended from a pole in the stern, out of which they could not climb.

I reached down for the octopus as it flowed past my feet, thinking to heft it up by one of its arms. It was slimy to the touch, slippery enough that it was virtually impossible to grip, and the arm flowed like water right through my hand. The octopus had no problem gripping me, however, and laid the sucker discs of another arm on the back of my groping hand—it was like being sniffed by an inquisitive dog. At the same time, its glance fell on me, and I gazed into its two pop eyes for an instant. I sensed a tightly controlled panic, an urgent desperation. But there was more there than that. I was jolted by the sense of an individual behind that look, a thinking entity. I have pulled many fish out of the water, but the look in that octopus's eyes was something else entirely. Some sort of visual conversation was going on between us, but it didn't last long. The octopus had no time for me. After a brief moment its gaze moved on, and it resumed its fruitless investigation of the boat, leaving faint round love marks on my skin where the suction discs had pulled my blood to the surface. The slipperiness left on my hand smelled of the sea, marine but not fishy.

A couple of minutes later, Pere Gras had no trouble whatsoever handling the octopus when its explorations brought it within arm's reach of where he was standing by the winch, winding in his clay pots. He wasted no time in bending down and grabbing the octopus with his left hand at the base of its head, hoisting it up. Suddenly, there was a short, wooden-handled knife in his right hand, which he plunged into the animal beneath its head at the juncture where its arms joined. The octopus instantly flopped limp in his hand, its search for escape ended. Gras cut out the beak and tossed it overboard, turned the entrails sac inside out into the sea, and washed the ink off. The octopus was dead and cleaned before its sucker marks faded from my arm, like the voice of someone who has died left over on an answering machine.

Pere Gras tossed the dead octopus into an ice chest on deck. "For my freezer," he said, smiling. "That's just the right eating size."

I was not alone in feeling the communicative nature of an octopus's gaze. Virtually anyone who has ever spent time with octopuses will say the same thing. The great oceanographer Jacques Cousteau wrote that an undersea diver will immediately feel something responsive looking into an octopus's eyes: "One has the sensation of lucidity, of a look much more expressive than that of any fish, or even any marine mammal."[1]

Octopuses are found on pendants, vases, and bowls from the Minoan civilization, as early as 1500 BC. Something more about the animal than just its taste has always called to the human eye as being special, worthy of consideration. Octopus images continued to be common decoration for the Greeks and Romans. The one constant in Greek representations is the animal's gaze, according to researchers in a 2018 article: "These octopus designs look straight at the viewer with two large eyes simultaneously recognizing both viewers and its existence. This exchange builds psychological dimension that includes both a sense of mutual recognition and self-awareness. The Ancient Greek artisans have successfully represented this rather abstract and enigmatic character of cephalopods and our metaphysical relationship with them."[2]

The octopus's otherworldly form is representative of the diverse biological forms we find in the world, and artists continue to find it worthy of special attention. Octopuses alone, or with other marine animals, are frequently found in the work of the extraordinary contemporary Spanish artist Miguel Barceló, from the Mediterranean island of Mallorca. He told the Spanish daily *El País*: "I have a lot of relationships with octopuses. I watch them for hours. They are amazing. They have memory, and their brains are capable of storing images. I try to learn from them."

On Mallorca, Barceló grew up hunting, killing, and eating octopuses. "I've hunted thousands, and now I don't kill them [anymore]," he told the newspaper. "I just about don't eat them either, because it's difficult to eat an animal with which you are so involved."[3]

As a result of my own contact with that inquisitive octopus gaze, I too have curbed my appetite for them out of respect, but it hasn't been easy. I first tasted octopus more than fifty years ago, in 1968, when I was living with two friends in a lean-to we constructed from canes and sticks on a deserted beach on the Greek island of Ios. At one end of the beach was a small café, the only business within walking distance. We were homesteading on the beach during tourism's offseason, and there was always supper to be had at the café for a more-than-reasonable price, ourselves the lone diners.

Octopus is prepared for the table in numerous ways around the world, but the first step before cooking it is always the same: It must be tenderized. For people with a freezer at home, tenderizing is not a problem. A week in the freezer is sufficient to break down muscle fiber and render an octopus ready to prepare. Of course, people have eaten octopus long before freezers existed. "The octopus must be beaten with twice seven blows" is an ancient Greek proverb.[4] One afternoon on that Ios beach I came across the café's proprietress, in her customary black dress, beating a dead, limp octopus against an outcrop of rocks along the shore, holding it by one of its eight arms and lifting it over her head, again and again, to smash it against the rocks, exactly as women before her had been doing for millennia. She gave me to understand that she was tenderizing it to serve for supper. It did, indeed, turn out to be plenty tender, as well as surprisingly tasty.

Grilled, seared on the outside, the meat we ate on Ios that evening was white, juicy, and delicious on the inside. I had a moment's hesitation before putting the first piece of thin-sliced suckered arm in my mouth, but only the first bite. After that, I added octopus to the list of things I dearly loved to eat, and I have done so in a number of countries, prepared in a variety of ways, all of them tasty. Fifty years later, the island of Ios is now a favorite watering hole for the world's well-to-do; the place has been developed nearly out of recognition. Ragtag young foreigners are no longer welcome to build lean-tos on the beach, and octopuses from the Aegean Sea, while ubiquitous on Greek island menus and still as tasty as ever, are priced too high for a backpacker's pocket.

Those people who have the misfortune to try and eat octopus that has not been sufficiently tenderized are not likely to try it a second time. In 1875, Henry Lee described being invited to dinner at a well-known doctor's house and served octopus, which the doctor's cook evidently did not know how to prepare: "Manfully and perseveringly for some minutes I tried to masticate a mouthful of it, but it was useless; and feeling that if human teeth could make no more impression on it than the sole of an old boot, the human stomach incurred risk of difficulties which all the well-known medical skill of our good host might be unable to cure, I declined to sacrifice myself to an idea, and, well, I did not swallow it."[5]

In the twenty-first-century Western world, octopus is often found in upscale eateries, prepared à la gourmet, but it still must be tenderized before cooked. The octopuses served at Manhattan's Michelin three-star restaurant

Le Bernardin were delivered, already tenderized, by the seafood market in Staten Island owned by Vincent Cutrone. Instead of beating them, he put the octopuses in converted washing machines in the back of his shop. "What I do in these tubs is the equivalent of beating an octopus on the rocks," he told me. "We defrost the octopuses, clean them, and put them in the machine. We cover them with water and add salt and ice. . . . Roughly a half hour later we'll change the water, and we'll do that until the new water stays clear and the octopus curls up beautifully."

That was just how Eric Ripert, Le Bernardin's superstar chef, wanted them to arrive at his kitchen. He liked to take a small octopus and serve it whole, curled up on a plate, he told me when we spoke at the restaurant. "At Le Bernardin, I always like to have a raw or marinated fish to begin with, and then something which is sometimes a little bit adventurous for the client, before they have something more substantial."

A small octopus fit the bill perfectly. Clients at Le Bernardin were paying plenty to address those octopuses head-on before chowing down on them. Occasionally someone sent an untouched octopus back to the kitchen, Eric Ripert told me, but it did not happen often. "I want to serve it as an octopus. To see the entire animal like that is slightly freaky, but at the same time it's a delight. If you're eating octopus you have to be able to handle it. It looks beautiful on the plate like that. . . . We're not going to slice it and hide its nature; we're saying, 'Look, it's right here.'"[6]

In 2025, the great preponderance of social media posts regarding farmed octopus comes down heavily on the side of prohibiting their aquaculture. Little attention has been paid to the advantages of eating them, including their superiority over other kinds of meat in terms of the effects on the environment or their high-quality protein and absence of fat.

In a 2024 article for the *New York Times*, Cara Buckley wrote, "Globally, food systems make up a third of planet-heating greenhouse gasses, with the environmental toll of the meat and dairy industries being particularly high. Livestock accounts for about a third of methane emissions, which have 80 times the warming power of carbon dioxide in the short term."[7]

The conversion rate of what an octopus eats into a high-protein, low-calorie meat is much better than any land-based livestock and higher than that for other popular aquacultured animals like salmon or shrimp. Octopuses can gain up to 2 percent of their body weight per day and can be frozen for up to three months with no significant loss of protein. Ultimately, climate change may mean that it will cost more to farm pigs, cows, or chickens than to grow cephalopods. Octopus is potentially a more environmentally sustainable farmed meat than beef, pork, or poultry, and in a world shaped by rising temperatures it could become a meat of choice.

Furthermore, they reproduce in great numbers and don't take long to do so. A cow may have one or two calves a year and a pig a litter or two annually, but while a female *Octopus vulgaris* lays only one clutch of eggs in her lifetime, she lays over a hundred thousand of them. Around 95 percent of those eggs will hatch and be eaten before the paralarvae reach two months old, devoured by any one of the numerous marine animals that feed on them. Farming would allow them to reach adulthood in great numbers. Researchers have shown that the hundred thousand eggs of *O. vulgaris* can be maintained and hatched using artificial incubators, with low light and correct seawater flow, without a female octopus present.[8]

Marine biologists posit that less than 5 percent of the hatched paralarvae transform into juveniles and settle into octopus life on the ocean floor. Even though in the wild newly hatched *O. maya* will head straight to the bottom, most of them, as well, will be eaten before becoming adults. If the paralarvae and juveniles from these species could, instead, be saved from the premature death that awaits them in the wild and grown to edible-sized octopuses on a farm under favorable conditions, they could both live out their lives—which would otherwise be cut drastically short—and serve to attenuate human hunger in many places.

In an era of global warming, it may become necessary to increase our consumption of environmentally sustainable foods as a key to saving us from a devastating food scarcity due to climate change. At the least, many of the foods we currently consume may become much more expensive because of weather. This may present a dilemma for the many environmental activists who loudly oppose octopus farming but who may be confronted with the fact that it is a more environmentally sustainable meat than what is currently consumed. If octopus eventually appears in quantity at the supermarket as a nutritious meal, and it's better for the planet than cows, pigs, or chickens, it could undermine arguments against octopus farming. And if octopus supplies increase and prices lower at the same time as the price of other meats rises, more people may be willing to include octopus in their diets or at least try it.

Yet in spite of their sustainable qualities, octopuses are a hard sell, even for those who are not opposed to eating them for ethical reasons. A number of studies have shown that there is a considerable gap between what people believe is environmentally responsible to eat and what they will actually consume. It is not easy to change eating habits, even when a person is convinced it would be good to do so. It might be salutary for the planet if I incorporated those Ecuadoran palm tree grubs or some other insect into my diet, but I have to admit that I'm not in a hurry to do so. A 2020 article in the journal *Frontiers in Psychology* recognized the problem:

> The challenge of convincing people to change their eating habits toward more environmentally sustainable food consumption patterns is becoming increasingly pressing. Food preferences, choices, and eating habits are notoriously hard to change as they are a central aspect of people's lifestyles and their socio-cultural environment. Many people already hold positive attitudes toward sustainable food, but the notable gap between favorable attitudes and actual purchase and consumption of more sustainable food products remains to be bridged.[9]

No surprise there. It also proves pretty hard for progressive-thinking environmentalists to give up their cars, their airline trips, and plenty of the other nonsustainable habits to which we in the Western world are accustomed. Yet more and more people in the United States and Europe are renouncing these things.

Should influencers, national commissions, and policymakers be tasked with promoting octopus as an acceptable food? Some 780 million people went hungry in 2022, up by 122 million from prepandemic 2019. Over three billion people on our planet cannot afford a healthy diet.[10] It may not be worth introducing another sentient being into the food chain aimed at an affluent market, thereby killing yet more animals for those who can already afford to buy whatever protein or meat sources they want. But perhaps, if octopus could help reduce the number of people in the world without access to a healthy diet, policymakers may want to look at how to introduce its widespread consumption.

In addition to considering whether large quantities of inexpensive farmed octopus might have the potential to ameliorate food insecurity, global warming may be creating an appropriate time to introduce octopus to our diets. The question of how climate change will affect what we eat is not yet known, though most researchers concur that in the long run our diets are bound to be affected to some degree. Some of the foods to which we are accustomed may no longer be sustainable or affordable, and this is particularly likely in the case of livestock. The willingness to try a new food may soon become urgent, maybe even as important to survival now as it was during the Inquisition in the sixteenth century, when not eating pork was enough to put your life at risk. It may be that incorporating new food into our diets will become less a matter of choice and more a matter of necessity.

How warming oceans will affect octopus populations is still uncertain, but early research is not encouraging. Studies of both *O. maya* and *O. vulgaris* indicate that warming waters could negatively affect fertilization of eggs, embryo development, and bring ruin on any fishery that is dependent on them. Climate change is going to affect the lives of many octopuses, which may in turn be a blow to the humans who fish them in the wild.

Carlos Rosas's research indicates that rising temperatures will reduce the inshore octopus population. When *Octopus maya* deserts the Yucatán coast for colder, more easterly waters, places too far out to fish with jimbas, he worries that many Sisaleños will be dropped into deep poverty. Likewise, artisanal fishers in Galicia, Portugal, and Dakhla are going to lose their source of income if octopuses become inaccessible to them. In a 2024 study, Carlos Rosas and his team at Sisal concluded that rising temperatures could have a negative effect on artisanal octopus fishers and result in an increased food insecurity in the Yucatán.[11]

His studies indicated that twenty-seven degrees Celsius is the maximum ocean warmth that *O. maya* can tolerate without being affected, Carlos Rosas told me in 2024, which was a temperature already being reached, and surpassed, on numerous summer days in Sisal. His research showed that above that temperature, reproductive behavior will be negatively affected. Rising temperatures reduced caloric intake and thereby reduced the energy disposable for reproduction, along with negatively affecting reproductive organs. "By 2050, there will be an increase of three degrees. What is now twenty-seven will be thirty. It's already happening. What's going to happen to the octopuses? They are not going to disappear entirely, but will move to cooler waters, further out, eliminating a source of income for thousands of families here.

"Another thing is that these octopuses come to within a few kilometers of the coast to reproduce. If the coastal water is too hot and they get here and lay their eggs, what's going to happen is [the eggs] will be deformed. Things are going to be horrible. Horrible! We did an experiment here last year to see what would be the effects of higher temperatures on the embryos. In their deformation, the embryos' ink mixed with its blood. Of the embryos maintained at thirty degrees, which were hatched by females who had been kept in thirty-degree water, 70 percent died. With elevated temperatures, those females who come to lay their eggs will not produce viable eggs."

Other researchers believe that octopuses are well designed for thriving in a changing climate: "The high plasticity of life history traits associated with short life spans and opportunistic feeding regimes are the key determinants allowing octopod and squid populations to respond quickly to changes in climate regimes. The fact that they 'live fast and die young' and act like 'weeds of the sea' may allow them to more quickly adapt to future global warming than their fish competitors and predators,"[12] wrote one group of leading octopus researchers in 2019.

In addition, climate change is going to affect the oxygen content of coastal waters and the strength and direction of ocean currents. Even if octopuses prove able to adapt to these changing conditions, it may be that those animals that make up the octopod diet are affected, which in turn may mean food scarcity for octopuses. In short, no one can yet say with certainty what effects warming oceans will have on octopuses, but early indications for some of the more commercially exploited species are concerning. While octopuses are, indeed, highly adaptable, a number of studies have confirmed Carlos Rosas's pessimistic predictions on the effects of warming temperatures on embryos, which appears to hold equally true for both *O. Maya* and *O. vulgaris*. A 2014 study of *O. vulgaris* concluded that when water temperature rose by three degrees Celsius, fewer embryos developed, and for those that did, the survival rate was significantly decreased.[13]

Salinity is one of the most important factors in what makes a good habitat for *O. vulgaris* and *O. maya*. As the planet's climate changes, some places will have heavier rainfall, while others will suffer droughts. Both of these will affect salinity. Perhaps the increased melted polar ice will combine with increased rainfall in some places to dilute the oceans and lower salinity. Or as melting ice caps increase water levels in oceans and seas, and areas experience longer dry seasons and more pervasive droughts, the salinity of the waters may increase.

Whatever the result of melting glaciers and ice caps, it is likely to have an effect of some kind on the octopuses we eat and to destabilize what is still, as of 2025, a relatively stable population. While the catch is generally declining, many *O. vulgaris*, *O. maya*, and *O. sinensis* still inhabit our oceans and seas. Most immediately in danger from warming waters are not octopuses but the people who fish for them and support their families by doing so.

The variations that could come with global warming might prove an economic advantage for the consistent production of aquaculture, Carlos Rosas told me, although he added that it makes him deeply sad to imagine Yucatán's artisanal octopus fishery wiped out. "Here [at the farm] we maintain an optimum temperature, and we can implement a strategy that may not produce three thousand tons like they want to do at Nueva Pescanova, but it can be an alternative source of income for the people on the coast, especially older people," he told me. "What's more, aquaculture is an educational tool: It helps transform the perspective people have of nature. It can change the way people exploit natural resources."

. . . .

For some people, eating octopus would be closer to consuming the family dog or cat than eating beef or pork. Accurate figures about how many people have octopuses as pets are nonexistent, but a rudimentary online search produces plenty of places where they can be bought as well as a number of groups exchanging information about keeping them in home aquariums.

Octopuses lack the qualities that most people look for in a pet: Many species are nocturnal; only one can be kept in a tank, and anything else alive placed in the tank is likely to be devoured; all octopuses have short lifespans; and substantial expense is involved when buying and maintaining the means to keep them alive for even that short time. To keep an octopus, say a popular choice like *O. bimaculoides*, the California two-spot octopus, requires a fifty-gallon saltwater aquarium with a good pump and filtration system, and those are not cheap.[14]

For the thrill factor, people also occasionally keep *Hapalochlaena maculosa*, a small brown octopus known as the blue-ringed octopus for the bright blue circles that appear on its skin when it is disturbed. One bite from this diminutive creature can deliver enough of the venom tetrodotoxin to paralyze ten adults or kill one person. It is the same venom contained in the puffer fish, which annually kills adventurous Japanese diners who order a dish of puffer fish called fugu and which chefs must be specially licensed to prepare. One company specializing in selling exotic marine animals for pets online will FedEx customers a blue-ringed octopus for $209.99.

While many exotic pets can be handled frequently, octopuses are not among them. To do so can be dangerous for both the animal, which can have its protective skin slime rubbed off, or the person doing the handling. While not frequent, octopuses do bite. I saw people working at the Sisal farm get bitten a couple of times. The bites are not deep and don't generally leave much of a wound, but they do break the skin, and they sting. Roy Caldwell, an octopus researcher at University of California, Berkeley, recorded his observations of a bite he got in the lab. "Reaction: bleeding, redness, other visible effects, pain . . . bleeding for about 10 minutes; bee sting like with redness and local swelling. It lasted about 2 hours, still sore the next day."[15]

That a market exists for pet octopuses is undeniable. And people do make livings in the exotic pet trade, so there is no reason to think that aquacultured octopuses might not eventually find a niche market waiting. Unfortunately for octopuses, they are much more frequently subjects in research laboratories than pets in an aquarium, and what is usually done to them in the lab is rarely pleasant.

The octopus is well suited to laboratory research. As long ago as 1971, the British behavioral psychologist G. D. Sanders, who did research on octopus behavior at the Stazione Anton Dohrn, urged researchers to make more use of them in their laboratories. His recommendations sound far from sanguine from the octopod point of view, and it is, perhaps, not surprising that following his retirement from academe he began writing thriller-mysteries featuring murder and mayhem. He has said those books allow him to be much more creative than when he was writing scientific articles. In his "The Cephalopods," wearing his academic hat, he wrote:

> Octopuses are voracious feeders and are not easily satiated so that motivation is rarely a problem. . . . Octopuses are particularly suitable for lesion studies as they withstand considerable surgical interference and recover rapidly. Small lesions are closed by contraction of the musculature and no special aseptic operating conditions are required. The higher centers of the brain are readily accessible through the soft cartilaginous cranium, in which the brain is enclosed.[16]

The European Union rules pertaining to the protection of lab animals are applicable to all vertebrates and some invertebrates including octopuses. An EU directive, passed in November 2010, stipulated that cephalopod research subjects must be treated exactly the same as mammalian subjects like monkeys or mice. This means that those researchers who are designing experiments with octopuses will need to have their work authorized by the national "competent authority" responsible for the implementation of the directive.[17] To gain approval, a project has to provide information that includes the relevance of the research for which the cephalopods are to be used; housing and conditions under which the animals will be cared for; the planned use of anesthesia and analgesia to reduce pain; the ways in which suffering will be reduced to a minimum; how the animals will be sacrificed if need be; and the qualifications of the project's personnel.

The EU's regulations only go so far in the effort to set standards to avoid unnecessary pain and suffering of lab animals and to provide for their safe accommodation. The rules allow considerable margins, both in what can be done in a laboratory to a living animal and when and how it can be sacrificed. Nevertheless, some regulations do exist, and the EU has stated that its ultimate goal is to eliminate the use of lab animals in all EU countries. While regulations may not make life in the laboratory an acceptable substitute for life in the wild, they are a step toward recognizing that we should take a minimum of care for those animals that we use to expand our own knowledge.

In the United States, while lawmakers in many places eagerly put forth legislation banning octopus farming, no state or federal protections exist for cephalopods used in research. In September 2023, the National Institutes of Health (NIH) proposed guidelines regarding the use of cephalopods in the laboratory and opened a period of comment until late December. This was in response to what was characterized as increased evidence that cephalopods feel pain and suffer from it.

No clear agreement exists among scientists as to whether many invertebrate animals experience pain, but it has become widely accepted that octopuses do. A number of experiments have been carried out, like one in 2020 in which octopuses received a subcutaneous injection of dilute (0.5%) acetic acid into one arm. Their subsequent reactions (grooming the affected site) and actions (avoiding the previously preferred space where the injections were given) strongly indicated that they had suffered pain from the injections.[18]

Notwithstanding the body of evidence showing that octopuses suffer pain, by March 2024, the NIH was backing away from proposing any new guidelines, citing a general lack of evidence about many cephalopods. "I'm sure from a legislative point of view it's easy to just group them in as cephalopods, but there's dozens and dozens of species, separated by millions of years of evolution, that need their own special standards to be set for them," said Connor Gibbons, a cephalopod facility manager in a research lab at Columbia University, in an interview.[19]

While the use of octopuses remains virtually unregulated in the United States, researchers are doing a lot of work with them in laboratories. They are seen by some as the next "model animal," becoming as common in labs as mice and rats. In fact, they are even being aquacultured for use in labs. At the Marine Biological Laboratory in Woods Hole, Massachusetts, on Cape Cod, researchers have successfully aquacultured a diminutive species for lab use. The Woods Hole facility, founded in 1930, is the United States' equivalent of the Stazione Anton Dohrn in Naples. An impressive amount of the work on octopus physiology and behavior has come from researchers affiliated with Woods Hole. Like the Stazione, it attracts many marine biologists as a place to work and pass their summers in a pleasant coastal research center.

In a 2021 paper, Woods Hole researchers announced that they had successfully cultivated the pygmy zebra octopus, *Octopus chierchiae*, and hailed it as the first time octopuses suitable for lab research had been aquacultured. The pygmy zebra octopus is a small creature, less than two inches long when mature, and of no use as food for humans. But it is one of few octopus species to lay egg clusters more than once in its lifetime, and its genome has been sequenced, making it well suited to laboratory research.

Woods Hole scientists have recently carried out a diverse group of experiments on larger octopuses, including dosing them with MDMA (ecstasy) to study the mechanics of sociability. Other research has focused on how octopuses taste by touch, with their suckers functioning as chemical receptors, and how they are able to regenerate lost limbs. "We're at a really exciting moment for working with [octopuses]," Caroline Albertin, an evolutionary developmental biologist in Woods Hole, told *Scientific American* in 2021. "There's just a vast ocean of research and questions that we need to explore."[20]

Another place where researchers dosed octopuses with MDMA is a laboratory at Johns Hopkins University. The lead researcher, Gül Dölen, specializes in understanding how psychedelics work and how they might serve to help people with PTSD, or who have suffered a stroke, to recover their ability to interact socially. She told an interviewer that her results with octopuses in a series of 2018 experiments had surprised her. "Octopuses are asocial, so the vast majority of the 300 or so species that we know about, they'll kill each other if you put them in the same tank. They're really asocial, so I really thought, 'There's no way this is gonna work.'"

She was wrong.

So the main thing we learned is that, despite the fact that these brains are very different from ours, that they respond to MDMA in basically the same way, and so even though an octopus is asocial, even though it doesn't have a cortex, or an amygdala, or a nucleus accumbens, or any of the other brain regions that we thought are so important for MDMA's ability to encode pro-social behavior, the octopuses, when we gave them MDMA, they went from being asocial and avoiding the chamber that had the other octopus in it, to being social and preferring to spend most of their time in that chamber.[21]

A number of researchers have applied the structure and flexibility of octopus arms to robotics. A 2020 press release from Harvard University announced one such successful project, a joint effort with Beihang University in Beijing, which developed "an octopus-inspired soft robotic arm that can grip, move, and manipulate a wide range of objects. . . . Researchers control the arm with two valves, one to apply pressure for bending the arm and one for a vacuum that engages the suckers. By changing the pressure and vacuum, the arm can attach to an object, wrap around it, carry it, and release it."[22]

And at the University of Illinois, the description of the Neurosciences Department's Octopus Project explains why studying octopuses is integral to these robotic designs:

> While originally created to study the development of octopus larvae, we are now one of seven laboratories, both biologists and engineers, who interact and collaborate on the problems of soft-bodied robotics. We focus on understanding the brain and behavior of an octopus, and we focus on intelligent motor control of how the animal controls a soft, flexible body. Among our goals is to create computational models that simulate the distributed intelligence within the octopus.

The $7.5 million funding for the project has come primarily from the Office of Naval Research for the development and building of a "Cyberoctopus." This will be a "software equivalent to the marine animal that will help the team understand and leverage its ability to conduct distributed inference and decision-making, its embodied control and intelligence, and its ability to learn new behavior quickly."[23]

. . . .

Farming octopuses for lab research might one day provide a living, but it is not likely to make a farmer rich. The potential large profits from octopus aquaculture are as a food product, and a lot of money is waiting to be made. By mid-2025, however, despite the high global demand, Moluscos del Mayab's new land in Sisal had yet to produce its first octopus for sale; Nueva Pescanova had, apparently, put their plans for an octopus farm on hold; and there was still no farmed octopus available in Japan.

Despite the current lack of successful octopus aquaculture, the questions of whether it should be farmed for the table, as well as for the lab, will eventually demand an answer. The market for octopus as an upscale food shows no signs of slowing down, and many traditional fishing grounds are registering declines in their populations. Science is unlikely to stop using octopus in research. These factors point to the real possibility that farmed octopus will have a place in our futures. However, a successful farm is going to require not only a high level of biological savvy regarding the octopus but also the ability to counteract the inevitable strong pushback from campaigners for animal rights.

For instance, the Washington State legislature passed a bill in 2024, signed into law by the governor, which specifically prohibited octopus farming in the state, even though a number of species were already being commercially aquacultured there, including oysters, clams, geoducks, crayfish, trout, salmon, and carp. The bill passed into law, although no one in Washington State was trying, or had ever tried, to farm octopuses.

The bill's primary sponsor, Representative Strom Peterson, declared, "Octopus farming leads to suffering and sickness for one of the more intelligent and feeling animals in our oceans. It can lead to huge environmental and ecological effects as well. Octopus farming is harmful to the animals and the environment and is unnecessary. It's time to move on."[24]

In a statement following the passage of the Washington state legislation, Jennifer Jacquet, who had moved on from New York University to a professorship in environmental science and policy at the University of Miami, applauded the legislation: "The decision is historic and visionary, and shows that industrial octopus production is not inevitable. We can stop the mass production of octopuses before it begins."[25]

California was eyeing a similar ban in 2024, and the Kanaloa Octopus Farm on the big Island of Hawai'i, behind the Kona International Airport, was shut down by the authorities in 2023. In addition to doing research on growing and keeping octopus, the Kanaloa farm was reported to be giving "petting" tours to an average of seventy visitors a day, at fifty dollars a head.[26] Animal rights groups in both Mexico and Japan are applying increasing pressure on local and national governments to ban octopus farming and any attempts to develop it.

The most potentially serious legal problem for would-be octopus farmers was SB4810, a bipartisan bill introduced in the US Senate by Sheldon Whitehouse (D-RI) and Lisa Murkowski (R-AK) in July 2024, which would not only prohibit the aquaculture of octopuses in the US but also the sale of farm-raised octopus from other countries. The bill stalled in committee, but in 2025 the two senators reintroduced it as SB1947.

Animal rights groups applauded the legislation. "Scientists have proven octopuses are complex, intelligent creatures who can feel a full range of emotions. Instead of exploiting them, we must protect this dynamic species who suffer terribly in confined settings," states Luke Allison, the legislative affairs manager for the Animal Legal Defense Fund. "[The fund] applauds Senator Whitehouse and Senator Murkowski for taking the first step to ensure the United States sets a global precedent for octopuses' welfare."[27]

Without the stateside market, there would be much less incentive to farm octopus. Neither of the senators' offices replied to my repeated requests to speak with someone about the legislation. Whitehouse's office sent me the press release from the bill's introduction, and Murkowski's office did not respond. The 2025 bill was, once again, referred to the Senate Committee on Commerce, Science, and Transportation, where it was apparently moribund. Even without reaching the Senate floor and a vote, the very fact of its introduction was enough to create a ripple of anxiety for potential octopus farmers around the world.

Even with all this strong opposition, it is a good bet that sooner or later, in one country or another, a viable octopus farm will appear. "There will be nearly 10 billion people on Earth by 2050—about 3 billion more mouths to feed than there were in 2010," according to the World Resources Institute. "As incomes rise, people will increasingly consume more resource-intensive, animal-based foods. At the same time, we urgently need to cut greenhouse gas (GHG) emissions from agricultural production and stop conversion of remaining forests to agricultural land."[28]

As so-called underdeveloped countries become more developed, the appetite for meat in these places increases. For people who have been too poor much of their lives to eat meat regularly, one of the first things they want when their household finances improve is to put more of it on their tables. Who are we to tell them not to do so, we who have been eating as much meat as we want all our lives? It might be better for all of us, and the planet, if everyone could choose farm-raised octopuses instead of beef, pork, or poultry, thereby avoiding the devastating consequences of increased livestock farming.

As wild octopuses grow scarcer, the price rises and marketing efforts are primarily aimed at an upscale, affluent market. These days, generally only the well-to-do eat octopus, and in a world where so many people are undernourished it seems absurd to use more resources farming food for the rich, particularly when the animal is one as sympathetic as an octopus. In 2019, the noted animal rights campaigner and philosopher Peter Singer wrote to oppose octopus aquaculture: "When it comes to subjecting millions of intelligent, sensitive animals, from a species never before domesticated or farmed, to industrial-scale captivity in order to increase the market for a luxury food . . . the arrogance with which we humans are behaving toward other animals is revealed in all its stark brutality."[29]

Octopuses are remarkably prolific. What if successful farming operations increased supply to a point where prices fell dramatically? Octopus farmers could reasonably assert that they are providing an opportunity for octopus paralarvae to grow up and enjoy life as adults, an opportunity which most of the hatchlings would not otherwise have, at the same time that they are making a nutritious, healthy meat widely available. Would it not be preferable to allow octopuses to grow to maturity in tolerable conditions and then provide protein to undernourished human beings?

The Canadian researcher Jennifer Mather, who has spent her professional life working with octopuses and affirming their individuality, responded to Peter Singer's objections to farming octopus in the same article: "I would like it if no one killed and ate the intelligent and fascinating octopuses that I work with, either caught in the wild or farmed in captivity. But I am a realist; people have to eat."[30]

If a sustainable feed could be developed that did not deplete wild stocks of fish or crustaceans, and the conditions were created under which octopuses could be farmed with the animal's welfare as a prime consideration, it might serve as a meat that environmentally aware, carnivorous consumers could eat in good conscience—a dietary middle ground between eating livestock and vegetarianism. In the same way that many people who want the protein that meat provides but do not want to eat beef, pork, or chicken have become so-called pescatarians, eating only fish and seafood, so such people might add farm-raised octopus to their diets.

When I read the London School of Economics' conclusion that "high welfare" farming of octopuses is impossible, I could not help but remember how tranquil and content those big octopuses in Nueva Pescanova's basement appeared. Perhaps octopuses could be farmed on a sufficiently large scale to help reduce global food insecurity without subjecting them to overcrowded tanks or other conditions that will cause them to suffer. It is not hard to monitor whether octopuses are stressed or unhappy. When they are unsatisfied with their living conditions, their behavior shows it. They become aggressive, cannibalistic, and may even chew off their own arms. These are clear signs of stressed octopuses, and we don't need to be able to talk to them to know if they are okay with their surroundings. What's more, if they are stressed, their meat will be of poor quality.

The Galician marine biologist Angel Guerra was convinced that octopus aquaculture was possible, he told me over the phone, but he added that anyone wanting to successfully farm octopuses would have to spend the time and money to provide them with stress-free lives in environments resembling their natural habitats. "The octopus is an animal with a well-developed brain, and it requires space, tranquility, and not to be crowded together," he said. "And they are picky eaters. If they are stressed the protein they produce is of bad quality; the meat is simply not good. You have to cultivate them in excellent conditions. And all of this requires a lot of effort and expense, but it will have to be done in order to succeed."

If and when octopus farming becomes a reality, consumers will have to decide if they want it gracing their tables, and lawmakers will have to consider whether and how to regulate the new aquaculture. Both octopuses and humans have had to develop a high degree of cognition and an impressive ability to adapt and survive in the conditions surrounding them. We need to acknowledge what is wonderful about octopuses and tailor our treatment of them accordingly.

Octopuses have brief but full lives. They inhabit those lives as sentient beings with an amazing capacity for learning and an innate vulnerability to pain and suffering, much like human beings and other farm animals. We should treat those octopus lives with the respect they deserve. We should think hard about whether we need to farm them at all, and if so, how it can best be done. And if we eventually farm octopuses, perhaps we should do so with the goal of reshaping the market so aquacultured octopuses provide the neediest among us with a healthy, tasty meat at the same time that we reduce livestock farming and the harm it causes to us and our planet.

Acknowledgments

A NYONE who reads this book will discern the huge debt it owes for its existence to the eminent marine biologist Dr. Carlos Rosas in Sisal, Mexico. He encouraged and educated me, and he and his wife María Eugenia Chimal Hernández offered friendship and kindness as well as expertise.

Special thanks also to the Robert B. Silvers Foundation, which helped finance my travels and research.

Many people helped me along the way, but any mistakes are, of course, entirely my responsibility. In addition to being indebted to everyone who was kind enough to sit for an interview, I owe thanks to many other people who helped, including Miriam at the Casa Álvarez guesthouse in Mérida and the members of the Moluscos del Mayab cooperative in Sisal, who kept me fed and laughing while they educated me: Silvia del Carmen Canul Pardenilla, María Virginia Narcisa Uicab Novelo, Julio Ernesto Sierra Delgado, Juana de la Cruz Maldonado Ek, Antonio Cob Reyes, and Bianca Daniela Hernández Hernández. Also many thanks to Ian Gleadall, Claudia Caamal-Monsreal, Sherri Judd, Elio & Compañía: Marta Martínez, Elio Mazilli, Carlos Bosch, Leopoldo Blume, and Edda Manrique; Susie and Carl Fehrmann; Chris Loew; Tim Maguire; George Cowdrey; Lucía Alcaraz; Jessica Delgado, Daniel Rivers, Kaya Delgado-Rivers; Elaine Maisner; and Jean Schweid.

Critical editorial care was provided by Catherine Hodorowicz, Tara Jordan, Erin Granville, and Alex Gergely. They did a terrific job.

As always, my helpmate Carmen Martínez Gómez was indispensable.

Notes

INTRODUCTION

1. Susannah Savage, "Global Farmed Fish Production Overtakes Wild Catch for First Time," *Financial Times*, June 7, 2024.

2. "FAO: Aquaculture in 2012 Will Supply More Than 50% of World's Fish Consumption," *MercoPress*, November 14, 2011, https://en.mercopress.com/2011/11/14/fao-aquaculture-in-2012- will-supply-more-than-50-of-world-s-fish-consumption.

3. Alessandro Camillo, "Farmed Fish Overtakes Wild Catch for the First Time: What That Means for the Consumer and the Environment," *Impakter*, June 14, 2024, https://impakter.com/farmed-fish-overtakes-wild-catch-for-the-first-time-what-that-means-for-the-consumer-and-the-environment/.

4. Jonathan Birch, Charlotte Burn, Alexandra Schnell, Heather Browning, and Andrew Crump, *Review of the Evidence of Sentience in Cephalopod Molluscs and Decapod Crustaceans* (London School of Economics and Political Science, November 2021), www.lse.ac.uk/News/News-Assets/PDFs/2021/Sentience-in-Cephalopod-Molluscs-and-Decapod-Crustaceans-Final-Report-November-2021.pdf.

5. Elene Lara, *Octopus Factory Farming: A Recipe for Disaster* (Compassion in World Farming International, October 2021), 20, www.ciwf.fr/media/7447174/ciwf_octopus-report-_21_aw_web_single_pages.pdf.

6. Gemma DiCarlo, "University of Oregon Research Offers a Window into How Octopuses See," *Think Out Loud*, Oregon Public Broadcasting, August 2, 2023, www.opb.org/article/2023/08/02/octopus-eyes-oregon/.

7. Eugene Linden, *The Octopus and the Orangutan* (Plume, 2003), 28.

CHAPTER 1

1. Warwick H. H. Sauer, Ian G. Gleadall, Nicola Downey-Breedt, et al., "World Octopus Fisheries," *Reviews in Fisheries Science and Aquaculture* 29, no. 3 (2019): 279–429, https://doi.org/10.1080/23308249.2019.1680603.

2. Fray Diego de Landa, *An Account of the Things of Yucatán: Written by the Bishop of Yucatán, Based on the Oral Traditions of the Ancient Mayas*, trans. David Castledine (Monclem Ediciones, 2000), 152.

3. Quoted in George Stuart and Gene Stuart, *The Mysterious Maya* (National Geographic Society, 1977), 39.

4. de Landa, *Account*, 70.

5. Gabriela Torres-Mazuera, "Formas cotidianas de participación política rural: el Procede en Yucatán," *Estudios Sociológicos* 32, no. 95 (May–August 2014), 302.

6. "Octopeptide-2," RS Synthesis, accessed October 3, 2025, https://rssynthesis.com/product/octapeptide-2/.

7. Stuart and Stuart, *Mysterious Maya*, 126.

8. Roger T. Hanlon and John B. Messenger, *Cephalopod Behaviour*, 2nd ed. (Cambridge University Press, 2018), 52.

9. Henry Lee, *The Octopus; or, The "DevilFish" of Fiction and of Fact* (Chapman and Hall, 1875), 24.

10. Sauer et al., "World Octopus Fisheries."

11. "Plan de manejo pesquero de pulpo en el Golfo de México y Mar Caribe," Instituto Mexicano de Investigacion en Pesca y Acuacultura Sustenable, Gobierno de México, March 28, 2014, p. 12, www.gob.mx/imipas/documentos/plan-de-manejo-pesquero-de-pulpo-en-el-golfo-de-mexico-y-mar-caribe.

CHAPTER 2

1. "Descendants of Milesius of Spain, King of Braganza, Father of the Irish Race, the High Kings of Ireland: First Generation," RootsWeb, ~cnoelldunc, "Medieval England Ireland Scotland and Wales," Noell-Turpin Family History Society, accessed October 4, 2025, http://freepages.rootsweb.com/~cnoelldunc/genealogy/Ancient/Heremon/D1.htm.

2. Rafael Bañón Díaz, "Historiografía del pulpo en Galicia," Badalnovas.com, November 7, 2021, https://badalnovas.com/2021/11/07/rafael-banon-diaz-historiografia-del-pulpo-en-galicia/.

3. Andrew Dalby, *Food in the Ancient World from A to Z* (Routledge, 2003), 236.

4. José Antonio Fidalgo Santamariña, *Polas rutas fo polbo e los polbeiros tradicionais en Ourense* (Deputación Provincial de Ourense, 2012), 37.

5. Elisa Lois, "Why Your Spanish Octopus Might Not Be Spanish After All," *El País*, December 29, 2017, https://english.elpais.com/elpais/2017/12/27/inenglish/1514377947_906602.html.

6. José Iglesias-Esteve, Francisco Javier Sánchez-Conde, and Juan José Otero-Pinzas, "Primeras experiencias sobre el cultivo integral del pulpo (*Octopus vulgaris* Cuvier) en el Instituto Español de Oceanografía," *Actas del VI Congreso Nacional de Acuicultura* (September–November 1997), Digital.CSIC, http://hdl.handle.net/10261/315426.

7. Nathan Allen and Guillermo Martinez, "World's First Octopus Farm Stirs Ethical Debate," Reuters, February 23, 2022, www.reuters.com/business/environment/worlds-first-octopus-farm-stirs-ethical-debate-2022-02-23/.

8. "Pescanova Analyzed the Galician Coast for the Construction of Its Octopus Macro Farm," *Tenerife Weekly*, December 7, 2023, https://tenerifeweekly.com/2023/12/07/pescanova-analyzed-the-galician-coast-for-the-construction-of-its-octopus-macro-farm/.

9. Jennifer Mather, "Daytime Activity of Juvenile *Octopus vulgaris* in Bermuda," *Malacologia* 29 (1988): 69–76.

10. European Market Observatory for Fisheries and Aquaculture Products, *Case Study: Octopus in the EU; Price Structure in the Supply Chain; Focus on Italy, Spain and Greece* (Publications Office of the European Union, 2021), https://data.europa.eu/doi/10.2771/87203.

11. "Octopus Marketing Report," Market Data Forecast, updated June 2025, www.marketdataforecast.com/market-reports/octopus-market.

12. Thomas de Cantimpré, *Liber de Natura Rerum*, 6.42, quoted in David Badke, "Octopus," The Medieval Bestiary (website), last updated September 24, 2023, https://bestiary.ca/beasts/beastsource105674.htm.

13. José Andrés Cornide, *Ensayo de una historia de los peces y otras producciones marinas de la costa de Galicia [. . .]* (facsimile ed., Edició do Castro, 1983), 184.

14. Victor Hugo, *The Toilers of the Sea*, vol. 2 (Little, Brown, 1888), 192–202.

15. Díaz, "Historiografía del pulpo."

16. F. Lishchenko, C. Perales-Raya, C. Barrett, et al., "A Review of Recent Studies on the Life History and Ecology of European Cephalopods with Emphasis on Species with the Greatest Commercial Fishery and Culture Potential," *Fisheries Research* 236 (April 2021), https://doi.org/10.1016/j.fishres.2020.105847.

17. Oxana Shamilyan, Ievgen Kabin, Zoya Dyka, et al., "Octopuses: Biological Facts and Technical Solutions," *2021 10th Mediterranean Conference on Embedded Computing (MECO)*, June 2021, https://arxiv.org/ftp/arxiv/papers/2201/2201.07885.pdf.

18. "Our Story," OctoNation, accessed September 19, 2025, https://shop.octonation.com/pages/about-us.

19. Charles R. Davis, "A Spanish Firm Wants to Kill One Million Octopuses a Year Using 'Ice Slurry' Baths at First-Ever Factory Farm," *Business Insider*, March 16, 2023, www.businessinsider.com/first-ever-proposed-octopus-farm-sparks-concern-over-conditions-2023-3.

20. Tik Root, "Inside the Race to Build the World's First Commercial Octopus Farm," *Time*, August 21, 2019, https://time.com/5657927/farm-raised-octopus/.

21. Davis, "Spanish Firm."

22. M. J. Wells, *Octopus: Physiology and Behaviour of an Advanced Invertebrate* (Chapman and Hall, 1978), 8.

23. John Z. Young, *The Anatomy of the Nervous System of Octopus vulgaris* (Oxford at the Clarendon Press, 1971), vii.

24. Graziano Fiorito, interview by the author, May 2010.

25. Graziano Fiorito and Pietro Scotto, "Observational Learning in *Octopus vulgaris*," *Science* 256, no. 5056 (April 24, 1992): 545–47.

26. Shamilyan, Kabin, Dyka, et al., "Octopuses: Biological Facts."

27. Jennifer A. Mather and Roland C. Anderson, "Exploration, Play, and Habituation in Octopuses (*Octopus dolfleini*)," *Journal of Comparative Psychology* 113, no. 3 (1999): 333–38.

28. "Genius Octopus Can Open Jars," clip from "The Octopus in My House," season 38, episode 3, of *Natural World*, aired on BBC Two on August 22, 2019, posted to YouTube by BBC Earth on September 25, 2021, www.youtube.com/watch?v=-KS-yI8VTfo.

29. Matt Shipman, "Aquaculture Does Little to Conserve Wild Fisheries, According to Study," *Phys News*, February 11, 2019, https://phys.org/news/2019-02-aquaculture-wild-fisheries.html.

30. Shipman, "Aquaculture Does Little."

31. Tristen Taylor, Ingrid Gercama, Nathalie Bertrams, and Ching-Li Tor, "Caught in Mauritania and Eaten in Japan, the Common Octopus Is Being Fished Out," *Daily Maverick*, November 7, 2023, https://pulitzercenter.org/stories/caught-mauritania-and-eaten-japan -common-octopus-being-fished-out.

32. Heather Carter and Scott Miller, "Business Bites: Spanish Seafood Giant Gets Ink in Its Eye over Octopus Farming Fiasco," Supply Side, April 15, 2024, www.supplysidefbj .com/food-beverage-operations/business-bites-spanish-seafood-giant-gets-ink-in -its-eye-over-octopus-farming-fiasco.

33. Xosé Manuel Beiras, El atraso economico de Galicia (El Edicions Xerais de Galicia, 1982): 129.

34. Quoted in Beiras, El atraso, 130–31.

35. "La actividad y financiación de Abanca suponen un impacto del 16% del PIB y el 15% del empleo en Galicia," Europa Press Galicia, January 15, 2025.

36. Servimedia, "Abanca espera 'batir' este año los resultados de 2023 y confía en que haya 'espacio' para negociar el impuesto a la banca," *La Vanguardia*, February 5, 2024, www .lavanguardia.com/economia/20240205/9513346/abanca-espera-batir-ano-resultados -2023-confia-haya-espacio-negociar-impuesto-banca-agenciaslv20240205.html.

37. Silvia Fernández, "La granja de pulpos de La Luz, en el aire por la situación financiera que arrastra Nueva Pescanova," *Canarias7*, October 1, 2023, www.canarias7.es/economia /granja-pulpos-luz-aire-situacion-financiera-arrastra-20231001224759-nt.html.

38. Luis Guitián Rivera, "La destrucción histórica del bosque en Galicia," in *Semata*, issue 13, *Historia Ecológica de Galicia*, eds. Luis Guitián Rivera and Augusto Pérez Alberti (Universidade de Santiago de Compostela, 2002), 107.

39. Miguel Anxo Murado, *Otro Idea de Galicia* (Debate, 2008), 15.

40. Santamariña, *Polas rutas fo*, 37.

41. Santamariña, *Polas rutas fo*, 42.

42. José Andrés Santiago Peieira and Juan José García del Hoyo, eds., *Observatorio Científico de las Pesquerías Artesanales: Recursos Pesqueros* (Servicio de Publicaciones de la Universidad de Huelva, 2021), 159.

43. Cristina Pita, Katina Roumbedakis, Teresa Fonseca, Fábio L. Matos, João Pereira, Sebastián Villasante, et al., "Fisheries for Common Octopus in Europe: Socioeconomic Importance and Management," *Fisheries Research* 235 (March 2021): 12, https://doi .org/10.1016/j.fishres.2020.105820.

44. Murado, *Otro Idea*, 40–41; "La historia del eucalipto en Galicia," blog post, Orballo, September 15, 2022, https://orballo.eu/historia-eucalipto-galicia/.

45. Rivera, "La destrucción histórica" 105–6.

46. Francesco De Augustinis, "Fish-Feed Industry Turns to Krill, with Unknown Effects on the Antarctic Ecosystem," *Mongabay*, October 19, 2022, https://news.mongabay .com/2022/10/fish-feed-industry-turns-to-krill-with-unknown-effects-on-the-antarctic -ecosystem/.

47. Lucia Baldi, Maria Teresa Trentinaglia, Massimo Peri, and L. A. Panzone, "Nudging the Acceptance of Insects-Fed Farmed Fish Among Mature Consumers," *Aquaculture Economics and Management* 28, no. 6 (2023): 1–32.

48. "Fish Waste Becomes Octopus Food as Farms Expand," *New York Post*, March 18, 2022, https://nypost.com/2022/03/18/fish-waste-become-octopus-food-as-farms-expand/.

49. David F. Willer, David C. Aldridge, Charlie Gough, and Kate Kincaid, "Small-Scale Octopus Fishery Operations Enable Environmentally and Socioeconomically Sustainable Sourcing of Nutrients Under Climate Change," *Nature Food* 4, no. 2 (2023): 179–89.

50. Eurogroup for Animals, "As Corporations Seek to Advance Breeding Efforts, EU Must Act to Stop Octopus Farming," press release, March 17, 2025, www.eurogroupforanimals.org /news/corporations-seek-advance-breeding-efforts-eu-must-act-stop-octopus-farming.

CHAPTER 3

1. Ole G. Mouritsen and Klavs Styrbaek, "Cephalopod Gastronomy—A Promise for the Future," *Frontiers in Communication* 3, no. 38 (2018): 1–11, https://doi:10.3389 /fcomm.2018.00038.

2. Ian Gleadall, "*Octopus sinensis* d'Orbigny, 1841 (Cephalopoda: Octopodidae): Valid Species Name for the Commercially Valuable East Asian Common Octopus," *J-Stage* 21, no. 1 (2016): 31–42.

3. Kouzo Itami, Yasuo Izawa, Saburo Maeda, and Kozo Nakai, "Notes on the Laboratory Culture of the Octopus Larvae," *Bulletin of the Japanese Society of Scientific Fisheries* 29, no. 6 (1963): 514–19.

4. Huaro Chiba, "Sea Farming Association Approves Octopus Farm," *SPC Fisheries Newsletter* 98, July–September 2001.

5. Warwick H. H. Sauer, Ian G. Gleadall, Nicola Downey-Breedt, et al., "World Octopus Fisheries," *Reviews in Fisheries Science and Aquaculture* 29, no. 3 (2019): 3, https://doi.org /10.1080/23308249.2019.1680603.

6. "Animal Welfare Campaigners Urge Japanese Government to Stop Octopus Factory Farming on World Octopus Day," Animal Rights Center, August 8, 2021, https://arcj.org /en/issues-en/animal-welfare-en/octopus-report/.

7. "Biology of Dicyemid Mesazoans," Hidetaka Furuya website, accessed November 8, 2025, www.bio.sci.osaka-u.ac.jp/~hfuruya/index2.htm.

8. Eric Nkando, "Sushi Conquers the World," *Fiestic Spark*, June 11, 2023, https://fiestic .com/spark/sushi-conquers-the-world/.

9. "Fukushima Octopuses Back on the Market," *Japan Times*, July 24, 2012, www .japantimes.co.jp/news/2012/07/24/business/fukushima-octopuses-back-on-the-market/.

10. Ian Gleadall, Goh Nishitani, Masami Abe, et al., "A Joint Academic-Industrial Project to Culture the Asian Common Octopus Commercially Using Land-Based Aquaria," International Council for the Exploration of the Sea, 2014, www.ices.dk/sites /pub/CM%20Doccuments/CM-2014/Theme%20Session%20P%20contributions/P2414 .pdf.

11. "Fully Farmed Octopus on Its Way to Your Dinner Table," *Nikkei Asia*, June 12, 2017, https://asia.nikkei.com/Business/Fully-farmed-octopus-on-its-way-to-your-dinner-table.

12. Chris Loew, "Following Profitable 2022, Nissui Unveils Its Future Aquaculture Goals," *SeafoodSource*, August 10, 2023, www.seafoodsource.com/news/premium/aquaculture /nissui-aquaculture-report-highlights-profitable-2022-unveils-company-s-future-goals.

13. Chris Loew, "Breakthrough Survival Rate Announced in Octopus Aquaculture Research," *SeafoodSource*, March 17, 2021, www.seafoodsource.com/news/aquaculture /breakthrough-survival-rate-announced-in-octopus-aquaculture-research.

14. Ricard Bru, "Tentacles of Love and Death: From Hokusai to Picasso," in *Imágenes Secretas: Picasso y la estampa erotica japonesa*, ed. Museu Picasso, Institut de Cultura de Barcelona (Barcelona Institute of Culture, 2009), 194.

15. Brassaï, *Conversaciones con Picasso* (Turner-Fondo de Cultura Económica, 2002), 275.

16. Raymond Lam, "Octopus Buddhas and Aliens: The Story of Tako Yakushi," *Buddhistdoor Global*, May 25, 2022, https://teahouse.buddhistdoor.net/octopus-buddhas-and-aliens-the-story-of-tako-yakushi/1000/.

CHAPTER 4

1. Anna Fleck, "The Key Players in Octopus Trade," Statista, June 29, 2023, www.statista.com/chart/30303/top-importers-and-exporters-of-octopus/.

2. Chris Loew, "Comprehensive View of World Octopus Fisheries Released," *SeafoodSource*, April 27, 2020, www.seafoodsource.com/news/supply-trade/comprehensive-review-of-world-octopus-fisheries-released.

3. Tristen Taylor, Ingrid Gercama, Nathalie Bertrams, and Ching-Li Tor, "Caught in Mauritania and Eaten in Japan, the Common Octopus Is Being Fished Out," *Daily Maverick*, November 7, 2023, https://pulitzercenter.org/stories/caught-mauritania-and-eaten-japan-common-octopus-being-fished-out.

4. Taylor et al., "Caught in Mauritania."

5. "Mauritania," Global Slavery Index, 2021, https://web.archive.org/web/20141023191039/http://www.globalslaveryindex.org/country/mauritania/.

6. World Bank Group, "Poverty and Equity Brief: Mauritania," October 2025, https://documents.worldbank.org/en/publication/documents-reports/documentdetail/099725504212516315.

7. John Miller, "Global Fishing Trade Depletes African Waters," *Wall Street Journal*, July 18, 2007.

8. "Sustainable Fisheries Partnership Agreement with Mauritania," European Commission, accessed November 10, 2025, https://oceans-and-fisheries.ec.europa.eu/fisheries/international-agreements/sustainable-fisheries-partnership-agreements-sfpas/mauritania_en.

9. Taylor et al, "Caught in Mauritania."

10. Donald J. Trump, "Proclamation on Recognizing the Sovereignty of the Kingdom of Morocco over the Western Sahara," The White House, December 10, 2020, https://trumpwhitehouse.archives.gov/presidential-actions/proclamation-recognizing-sovereignty-kingdom-morocco-western-sahara/.

11. Latifa Babas, "Polisario Alarmed over Potential U.S. Consulate in Dakhla Under Trump," *Yabiladi*, November 18, 2024, https://en.yabiladi.com/articles/details/156347/polisario-alarmed-over-potential-consulate.html.

12. ClientEarth, *Trade in Troubled Waters for the Octopus* (ClientEarth, July 2023), www.clientearth.org/media/13gfvuvd/trade-in-troubled-waters-for-the-octopus.pdf.

13. "One of the Richest Coastlines in the World," Western Sahara Resource Watch, May 11, 2023, https://wsrw.org/en/news/the-fishing-industry.

14. Victoria Veguilla, "How the Fishing Industry Strengthened Morocco's Occupation of Western Sahara," MEIRP, May 11, 2022, https://merip.org/2022/05/how-the-fishing-industry-strengthened-moroccos-occupation-of-western-sahara/.

15. Mayuka Tanabe, "The Green March Brings Forth the Desert Treasures: Japanese Cooperation and Morocco's South-Atlantic Fishing," *The Newsletter*, International Institute for Asian Studies, Autumn 2016, www.iias.asia/the-newsletter/article/green-march-brings-forth-desert-treasures-japanese-cooperation-moroccos-south.

16. Veguilla, "How the Fishing Industry."

17. Daryl Austin, "Sushi Is More Popular Than Ever, but Is It Healthy? What to Know," *USA Today*, May 19, 2023, https://eu.usatoday.com/story/life/health-wellness/2023/05/19/is-sushi-healthy-healthiest-rolls-weight-loss-benefits-explained/70212566007/.

18. Ian Gleadall, "Towards Global Sustainability for Cephalopod Seafood," *Marine Biology* 171 (2024): 44.

19. Elisa Lois, "Why Your Spanish Octopus Might Not Be Spanish After All," *El País*, December 29, 2017, https://english.elpais.com/elpais/2017/12/27/inenglish/1514377947_906602.html.

CHAPTER 5

1. Elise Titia Gieling, Rebecca Elizabeth Nordquist, and Franz Josef van der Staay, "Assessing Learning and Memory in Pigs," *Animal Cognition Aims and Scope* 14, no. 2 (January 2011): 151–73.

2. Evan Malmgren, "The Intelligence of Swine," *Noema*, November 24, 2021, www.noemamag.com/the-intelligence-of-swine/.

3. Lori Marino and Christina M. Colvin, "Thinking Pigs: A Comparative Review of Cognition, Emotion, and Personality in *Sus domesticus*," *International Journal of Comparative Psychology* 28 (2015), https://escholarship.org/uc/item/8sx4s79c.

4. Massimo Montanari, *The Culture of Food* (Blackwell, 1996), 3–13.

5. Giuseppe Santisi, Paola Magnano, and Grazia Di Marco, "Psychological Sustainability and Attitudes of New Food Consumption. A Research on Food Disgust and Neophobia," *Quality—Access to Success* 20 (2019): 549–55, ProQuest, https://www.proquest.com/docview/2198414266.

6. Paul Rozin, Jonathan Haidt, Clark McCauley, and Sumio Imada, "Disgust: Preadaptation and the Cultural Evolution of a Food-Based Emotion," in *Food Preferences and Taste: Continuity and Change*, ed. Helen Macbeth (Berghahn Books, 1997), 67.

7. Paul Fieldhouse, *Food and Nutrition: Customs and Culture* (Stanley Thomas Publishers, 1998), 194.

8. Mary Ann Bass, Lucille Wakefield, and Kathryn Kolasa, *Community Nutrition and Individual Food Behavior* (Burgess, 1979), quoted in Patricia Aguirre, "El carácter social de la alimentación," *Medicina y sociedad* (blog), August 2016, https://medicinaysociedad.wordpress.com/wp-content/uploads/2016/08/aguirre.pdf.

9. Patricia Pliner and Karen Hobden, "Development of a Scale to Measure the Trait of Food Neophobia in Humans," *Appetite* 19, no. 2 (October 1992): 105–20.

10. Norge W. Jerome, "On Determining Food Patterns of Urban Dwellers," in *Gastonomy: The Anthropology of Food and Food Habits*, ed. Margaret L. Arnott (Moulton Publishers, 1975), 100.

11. Hannah K. Bradshaw, Summer Mengelkoch, Matthew Espinosa, Alex Darrell, and Sarah E. Hill, "You Are What You (Are Willing To) Eat: Willingness to Try New Foods Impacts Perceptions of Sexual Unrestrictedness and Desirability," *Personality and Individual Differences* 182 (November 2021), 111082.

12. Carl E. Guthe, "History of the Committee on Food Habits," in *The Problem of Changing Food Habits: Report of the Committee on Food Habits 1941–1943*, National Research Council Committee on Food Habits (National Academies Press, 1943), www.ncbi.nlm.nih.gov /books/NBK224361/.

13. Margaret Mead, "Introduction: The Problem of Changing Food Habits," in *The Problem of Changing Food Habits: Report of the Committee on Food Habits 1941–1943*, National Research Council Committee on Food Habits (National Academies Press, 1943), https:// www.ncbi.nlm.nih.gov/books/NBK224361/.

14. Kurt Lewin, "Forces Behind Food Habits and Methods of Change," in *The Problem of Changing Food Habits: Report of the Committee on Food Habits 1941–1943*, National Research Council Committee on Food Habits (National Academies Press, 1943), https://www.ncbi .nlm.nih.gov/books/NBK224347/.

15. Ellen Messer, "Three Centuries of Changing European Tastes for the Potato," in *Food Preferences and Tastes: Continuity and Change*, ed. Helen Macbeth (Berghahn Books, 1997), 108.

16. Montanari, *Culture of Food*, 86.

17. Douglas Frantz and Catherine Collins, "3 Reasons to Avoid Farmed Salmon," *Time*, July 21, 2022, https://time.com/6199237/is-farmed-salmon-healthy-sustainable/.

18. "Looking Back at the Shrimp Market in 2022," *Seafoodnews*, December 27, 2022, www.seafoodnews.com/Story/1242492/Year-in-Review-Looking-Back-at-the-Shrimp -Market-in-2022.

19. "Farmed Shrimp," World Wildlife Organization, www.worldwildlife.org/our-work /oceans/sustainable-seafood/farmed-seafood/farmed-shrimp/.

20. Santiago Otamendi, interview by the author, March 2003.

21. William Bradford, *Of Plymouth Plantation:1620–1647* (Random House, 1981), 144.

22. Luis Villaverde, *Mariscos de Galicia* (Ediciones de Castro, 1974), 178.

CHAPTER 6

1. Jacques-Yves Cousteau and Philippe Diolé, *Octopus and Squid: The Soft Intelligence* (Doubleday, 1973), 88.

2. Riuka Nakajima, Shuichi Shigeno, Letizia Zullo, Fabio De Sio, and Markus R. Schmidt, "Cephalopods Between Science, Art, and Engineering: A Contemporary Synthesis," *Frontiers in Communication* 3, no. 20 (June 2018), https://doi:10.3389/fcomm.2018 .00020.

3. Karmentxu Marín, "Tengo Mucha Relación con los Pulpos," *El País*, May 1, 2011, 64.

4. Andrew Dalby, *Food in the Ancient World from A to Z* (Routledge, 2003), 236.

5. Henry Lee, *The Octopus; or, The "Devil Fish" of Fiction and of Fact* (Chapman and Hall, 1875), 88–89.

6. Eric Ripert, interview by the author, February 2007.

7. Cara Buckley, "These Cities Aren't Banning Meat. They Just Want You to Eat More Plants," *New York Times*, February 28, 2024.

8. Astrid Deryckere, Ruth Styfhals, Erica A. G. Vidal, Eduardo Almansa, and Eve Seuntjens, "A Practical Staging Atlas to Study Embryonic Development of *Octopus vulgaris* Under Controlled Laboratory Conditions," *BMC Developmental Biology* 20, no. 7 (2020), https://doi.org/10.1186/s12861-020-00212-6.

9. Iris Vermeir et al., "Environmentally Sustainable Food Consumption: A Review and Research Agenda from a Goal-Directed Perspective," *Frontiers in Psychology* 11, no. 1603 (July 2020), https://doi: 10.3389/fpsyg.2020.01603.

10. "Global Issues: Food," United Nations, accessed November 27, 2025, www.un.org/en/global-issues/food.

11. Ángel Escamilla-Aké, Luis Enrique Angeles-Gonzalez, Alejandro Kurczyn, Claudia Caamal-Monsreal, and Carlos Rosas, "How to Quantify the Regional Effects of Ocean Temperature Rise Due to Climate Change: Implications of *Octopus maya* Ecophysiology on Food Security of the Yucatán Shelf Artisanal Fishermen," *Regional Environmental Change* 24 (2024), article 81, https://doi.org/10.1007/s10113-024-02236-1.

12. Warwick H. H. Sauer, Ian G. Gleadall, Nicola Downey-Breedt, et al., "World Octopus Fisheries," *Reviews in Fisheries Science and Aquaculture* 29, no. 3 (2019): 17, https://doi.org/10.1080/23308249.2019.1680603.

13. Tiago Repolho et al., "Developmental and Physiological Challenges of Octopus (*Octopus vulgaris*) Early Life Stages Under Ocean Warming," *Journal of Comparative Physiology B* 184, no. 1 (2014): 55–64.

14. Nancy King, "Octopus Pet Checklist: Before You Get an Octopus as a Pet," TONMO.com, March 8, 2019, https://tonmo.com/articles/octopus-pet-checklist-before-you-get-an-octopus-as-a-pet.3/.

15. Roy Caldwell "Octopus Bites," TONMO.com, August 16, 2007 (page discontinued).

16. G. D. Sanders, "The Cephalopods," in *Invertebrate Learning*, vol. 3, *Cephalopods and Echinoderms*, eds. W. C. Coming, J. A. Dyal, and A. O. D. Willows (Springer, 1975), 2.

17. "Protection of Laboratory Animals," European Union, accessed December 13, 2025, https://eur-lex.europa.eu/EN/legal-content/summary/protection-of-laboratory-animals.html.

18. Robyn J. Crook, "Behavioral and Neurophysiological Evidence Suggests Affective Pain Experience in Octopus," *iScience* 24, no. 3 (March 19, 2021), www.cell.com/iscience/fulltext/S2589-0042(21)00197-8.

19. Calli McMurray, "Knowledge Gaps in Cephalopod Care Could Stall Welfare Standards," *Transmitter*, March 13, 2024, www.thetransmitter.org/policy/knowledge-gaps-in-cephalopod-care-could-stall-welfare-standards/.

20. Rachel Nuwer, "An Octopus Could Be the Next Model Organism," *Scientific American*, March 1, 2021, www.scientificamerican.com/article/an-octopus-could-be-the-next-model-organism/.

21. Gül Dölen, interview by Gideon Litchfield and Lauren Goode, "To Understand the Human Brain, Give an Octopus MDMA," *Have a Nice Future* podcast, *Wired*, July 12, 2023, www.wired.com/story/have-a-nice-future-podcast-13/.

22. Leah Burrows, "The Tentacle Bot," Harvard School of Engineering and Applied Sciences, February 27, 2020, https://seas.harvard.edu/news/2020/02/tentacle-bot.

23. "The Octopus Project," Neuroscience Program, University of Illinois Urbana-Champaign, accessed November 28, 2025, https://neuroscience.illinois.edu/octopus-project.

24. Julie Cappiello, "Octopus Farms: Washington State Ready to Ban, California Introduces Legislation," World Animal Protection, March 18, 2024, www.worldanimalprotection.us/latest/blogs/octopus-farms-washington-state-ready-to-ban-california-introduces-legislation/.

25. "Washington State Prohibits Octopus Farming: A Major Victory for Animals," Aquatic Life Institute, accessed November 13, 2025, www.ali.fish/blog/washington-state -prohibits-octopus-farming-a-major-victory-for-animals.

26. Clare Hamlett, "What Really Happened at That Controversial Octopus Farm in Hawaii," *Sentient Science*, August 22, 2023, https://sentientmedia.org/hawaii-octopus -farm-controversy/.

27. Meaghan McCabe, "Whitehouse, Murkowski Introduce Bipartisan Bill Banning Commercial Octopus Farming," press release, Senator Sheldon Whitehouse website, July 25, 2024, www.whitehouse.senate.gov/news/release/whitehouse-murkowski-introduce -bipartisan-bill-banning-commercial-octopus-farming/.

28. Janet Ranganathan, Richard Waite, Tim Searchinger, and Craig Hanson, "How to Sustainably Feed 10 Billion People by 2050 in 21 Charts," World Resources Institute, December 5, 2018, www.wri.org/insights/how-sustainably-feed-10-billion-people-2050-21-charts.

29. Peter Singer, response to "The Case Against Octopus Farming," *Issues in Science and Technology* 35, no. 3 (Spring 2019), https://issues.org/octopus-farming/.

30. Jennifer Mather, response to "The Case Against Octopus Farming," *Issues in Science and Technology* 35, no. 3 (spring 2019), https://issues.org/octopus-farming/.